Emerging Contaminants and Associated Treatment Technologies

Emerging Contaminants and Associated Treatment Technologies focuses on contaminant matrices (air, land, water, soil, sediment), the nature of pollutants (emerging, well-known, persistent, e-waste, nanomaterials, etc.), health effects (e.g., toxicology, occupational health, infectious diseases, cancer), treatment technologies (bioremediation, sustainable waste management, low cost technologies), and issues related to economic development and policy. The book series includes current, comprehensive texts on critical national and regional environmental issues of emerging contaminants useful to scientists in academia, industry, planners, policy makers and governments from diverse disciplines. The knowledge captured in this series will assist in understanding, maintaining and improving the biosphere in which we live. The scope of the series includes monographs, professional books and graduate textbooks, edited volumes and books devoted to supporting education on environmental pollution at the graduate and post-graduate levels.

Hassan Alshemmari
Muhammad Zaffar Hashmi
Yasar N. Kavil • Wang Shu-hong
Editors

Contaminated Land and Water

Remediation and Management

Springer

Editors
Hassan Alshemmari
Kuwait Institute for Scientific Research
Kuwait City, Kuwait

Yasar N. Kavil
King Abdulaziz University
Jeddah, Saudi Arabia

Muhammad Zaffar Hashmi
Institute of Molecular Biology
and Biotechnlogy
The University of Lahore
Lahore, Pakistan

Wang Shu-hong
College of Resource & Civil Engineering
Northeastern University
Shenyang, China

ISSN 2524-6402 ISSN 2524-6410 (electronic)
Emerging Contaminants and Associated Treatment Technologies
ISBN 978-3-031-65128-1 ISBN 978-3-031-65129-8 (eBook)
https://doi.org/10.1007/978-3-031-65129-8

This Springer imprint is published by the registered company Springer Nature Switzerland AG
The registered company address is: Gewerbestrasse 11, 6330 Cham, Switzerland

Preface

Over the years, human activities have left an indelible mark on our environment. Industrial processes, agricultural practices, improper waste disposal, and a myriad of other anthropogenic activities have led to the accumulation of hazardous substances in our soil and water. The consequences of this contamination are far-reaching, impacting ecosystems, human health, and even our economies. As the awareness of these issues grows, so does the urgency to address and mitigate the adverse effects of contaminated land and water. Fortunately, we have made significant progress in understanding the science of contamination, devising remediation techniques, and developing effective management practices. "Contaminated Land/Water, Remediation and Management" is a result of the collaborative efforts of numerous experts, researchers, and practitioners in the field of environmental science and engineering. This book aims to consolidate their knowledge and experiences to provide you, the reader, with a comprehensive and up-to-date resource on this critical topic.

Key features of this book include:

1. *An In-Depth Exploration of Contaminants*: We delve into a wide range of contaminants that can pollute our land and water, including heavy metals, organic compounds, pesticides, and emerging contaminants. Each chapter provides insights into their sources, behavior, and potential impacts on the environment and human health.
2. *Remediation Technologies*: Throughout the book, we explore various remediation techniques used to restore contaminated sites to a more sustainable state. From traditional methods like excavation and soil washing to innovative approaches like bioremediation and phytoremediation, you will gain a comprehensive understanding of the available tools for tackling contamination.
3. *Risk Assessment and Management*: Understanding the risks associated with contaminated sites is crucial for effective decision-making. We examine risk assessment methodologies and explore strategies for managing contaminated areas to protect both human populations and ecosystems.

4. *Legal and Regulatory Frameworks*: Navigating the legal and regulatory aspects of contaminated land and water management is a complex endeavor. Our book sheds light on the international and regional frameworks that guide pollution control and management efforts.
5. *Case Studies*: Real-world examples bring theory into practice. Throughout the book, we present case studies that highlight successful remediation projects, as well as challenges faced and lessons learned.

Kuwait City, Kuwait — Hassan Alshemmari
Islamabad, Pakistan — Muhammad Zaffar Hashmi
Jeddah, Saudi Arabia — Yasar N. Kavil
Shenyang, China — Wang Shu-hong

Contents

Chapter 1
Ecotoxicology and Risk Assessment of Contaminated Water

Jacky Bhagat, Nisha Singh, and Yasuhito Shimada

1.1 Introduction

The ever-increasing world population, anthropogenic activity, and industrial development have pushed the interaction boundaries of human society with science and technology. In the twentieth century, the advent of emerging water contaminants stimulated vast research focusing on the effects of water contaminants on different levels of biological organizations. Flooding reports of the occurrence of ubiquitous plastics and other emerging contaminants in the natural environment have kept ecotoxicologists busy investigating the impact of contaminants on aquatic organisms and humans. Responsible and sustainable development can be pushed forward to enhance the scientific field in such a way that human society has minimal disturbance to their well-being and environment. The complex matrix of the modern evolution of humans and environment management falls under the radar of several investigations. Ecotoxicology and risk assessment are efficient tools to predict and prevent inevitable ecological impact due to human actions.

Ecotoxicology is a rapidly evolving transdisciplinary field of research that integrates ecology, toxicology, physiology, environmental chemistry, and molecular

J. Bhagat (✉)
Graduate School of Regional Innovation Studies, Mie University, Tsu, Japan

Mie University Zebrafish Researc Center, Tsu, Japan

N. Singh
Japan Agency for Marine-Earth Science and Technology (JAMSTEC), Yokosuka, Kanagawa, Japan

Y. Shimada
Mie University Zebrafish Researc Center, Tsu, Japan

Department of Integrative Pharmacology, Mie University Graduate School of Medicine, Tsu, Japan

H. Alshemmari et al. (eds.), *Contaminated Land and Water*, Emerging Contaminants and Associated Treatment Technologies, https://doi.org/10.1007/978-3-031-65129-8_1

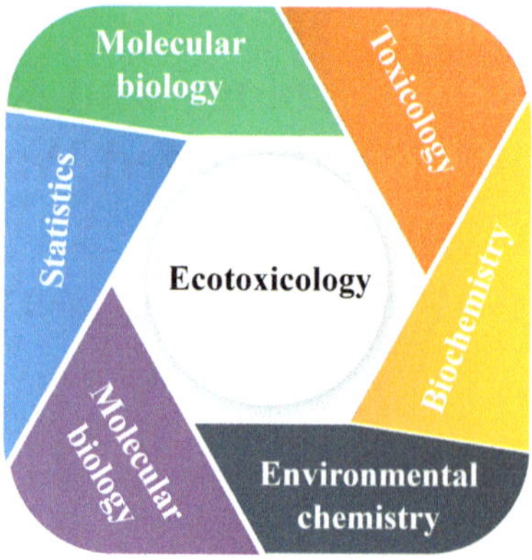

Fig. 1.1 Ecotoxicology—a transdisciplinary field of research

biology to name a few (Fig. 1.1). In essence, ecotoxicology focuses on two major areas: ecology, the study of the environment, and toxicology, which deals with the effect of contaminants on living organisms. The term "ecotoxicology" was first used in 1969 by a French researcher named René Truhaut, as a branch of toxicology that deals with the study of adverse effects of chemical pollutants on the ecosystem and its mechanisms to produce toxicity (Walker et al., 1996). Later in 1962, the book *Silent Spring* by Richard Carson proved a revelation and stimulated research toward measuring the effects of pollutants. Compared to other branches of science, ecotoxicology, a prominent research area, had a short gestation period. Ecotoxicological studies play an important role in monitoring contaminant levels and helping develop strategies to control pollution. It also includes the accumulation of contaminants in organisms and their transfer across trophic levels.

Risk assessment is a critical step to evaluate inherent hazards due to the transport or transformation of chemical contaminants with their exposure to the environment. Risk assessment of chemicals in the environment is a conceptual tool to estimate the probability that certain contaminants may cause deleterious impacts. Risk assessment uses data organization and analysis of the preexisting scientific knowledge and identifying a course of action. A robust risk assessment involves several steps which are divided into three broad categories: pinning down a problem, recognizing the components to be valued, and communicating and agreements among the stakeholders for better decision-making. The development of analytical techniques to measure the concentration of contaminants during the 1950s advanced research focusing on environmental contaminants. Measurement of mere chemical concentration in the environment is not sufficient to understand the overall pollution status of the ecosystem. In addition to the chemical concentration, the biological response of organisms in response to the contaminants is required for understanding the health status. Ecotoxicological studies include a range of techniques and tests to investigate the molecular mechanism underlining the toxic effects of environmental contaminants such as heavy metals, pesticides, plastics, organic contaminants, pharmaceutical and personal care products, etc.

Several review articles are available on the ecotoxicology of contaminants in water (Ali et al., 2019; Gworek et al., 2019; Li et al., 2019; Medjdoub et al., 2020; Valavanidis & Vlachogianni, 2010). Several book chapters have also been written focusing on the ecotoxicological effects of aquatic contaminants (Blasco et al., 2016; Blasco & Corsi, 2019; Kendall & Smith, 2006). The purpose of this chapter is to provide up-to-date information on ecotoxicology and risk assessment studies carried out on aquatic contaminants. First, the global status of water contamination along with sources and types of contaminants has been described. Later, the commonly used ecotoxicological biomarkers and indicator species have been mentioned. Finally, the concept of environmental hazards, risk assessment, and management has been discussed.

1.2 Ecotoxicological Research

There is a strong scientific consensus on the degradation of the environment due to an increase in anthropogenic activities. Ecotoxicological studies have become mandatory in industries for the production and use of industrial and agrochemicals. Strict regulatory frameworks and the growing awareness of the adverse effects of contaminants have stimulated ecotoxicological research around the globe. Research publications on ecotoxicological studies has been consistently growing over the years (Fig. 1.2).

1.3 Global Water Contamination

Water contamination is one of the major challenges faced by countries around the globe. It is widespread and has been a challenging problem in every country, irrespective of its developmental status. Contamination of water is a crucial global issue that humanity is facing in the twenty-first century. Thousands of new chemical products are produced and released every year into the aquatic environment. In the majority of countries, wastewater treatment plants are not efficient for the removal of toxic microcontaminants; hence, a large number of toxic contaminants enter the agricultural field through the usage of biosolids. Moreover, increasing evidence of illegal dumping of toxic chemical contaminants into the water bodies degrades the water quality and causes a serious threat to the inhabited aquatic organisms. Considering the fact that there is more than 1 lakh registered chemical for industrial and daily use, the severity of water pollution is evident (Schwarzman & Wilson, 2009). It is estimated that more than two million tons of waste are released every day into the water system (Norrrahim et al., 2021). Drinkable water resources are less than 1% of the total water supply. According to an estimate by United Nations, around 2.7 billion people will face a water crisis by 2025. Water contamination poses a serious risk to human health. Contaminated water can transmit waterborne year (Sharma et al., 2019). Over time, these pesticides enter the river through run-offs or are also leached into the groundwater. Extensive usage of pesticides and fertilizers is the major reason for agricultural contamination. In high-income countries, agricultural contamination has overtaken industrial contamination. In countries like the USA and China, agricultural contamination is responsible for the majority of surface water contamination. Nitrate is the most common chemical contaminant in agriculture and is responsible for the well-known blue baby syndrome. The last two decades have seen the emergence of several new contaminants such as antibiotics, growth promoters, vaccines, and hormones.

(c) **Industrial Effluents**

Industrial activities produce a large number of waste products that are often discharged into the natural water with limited treatment or left untreated. The lack of strict pollution control policies and their poor implementation in countries are major reasons for the increase in industrial pollution. Moreover, to avoid the high cost, industries often stick with obsolete technologies that produce more waste than modern technologies. A plethora of hazardous materials are used depending on the nature of industries that contaminates surface water and groundwater. Major contaminants from industrial effluents include heavy metals, nitrates, phosphates, oil, and petrochemicals. Accumulation of heavy metals in organisms living in the industrial effluent-polluted area shows the severity of pollution status (Ndimele et al., 2017; Zafarzadeh & Mehdinejad, 2015).

(d) **Aquaculture**

Aquaculture has grown rapidly in many coastal countries to meet the up-food demand. Since the 1980s, the aquaculture industry has grown 20-fold (Mateo-Sagasta et al., 2017). Aquaculture facilities often discharge a large amount of solid waste, chemicals, and therapeutics into natural waters. The effluent discharged from aquaculture farms is responsible for eutrophication (Chislock et al., 2013). Aquaculture can also affect the physicochemical properties of groundwater (Putheti et al., 2008).

(e) **Mining Operations**

Mining plays a vital role in economic development in several countries. It also poses a critical menace to ecosystem health and negatively affects terrestrial and aquatic life. Mining activities generate millions of gallons of wastewater, and incorrect disposal of water containing heavy metals and chemicals can have a devastating effect on the flora and fauna. Acid mining drainage, heavy metal contamination, erosion, and sedimentation are major problems associated with mining activities. Prolong mining activities can significantly deplete surface water and groundwater supplies. Several studies have reported surface water and groundwater contamination around the mining regions (Bud et al., 2007; Jhariya et al., 2016).

1.5 Ecotoxicological Contaminants

The modern world uses a variety of chemicals every day for industrial purposes, agriculture, and medicines and to improve the quality of life. Hundreds of thousands of chemical products are available in the market. The rapid increase in the production of such chemicals increases the risk of contamination in the environment. Industrial effluents contain a variety of toxic metals. Untreated release of effluent causes heavy metals and other contaminants to get deposited into the sediments. Plants and animals residing near the affluent area can take up heavy metals and also transfer them to higher trophic-level organisms. Accumulation of chromium and cadmium in plants and its repercussion on human health have been reported (Kapoor et al., 2022; Roy, 2019). (Pereira et al., 2009) have highlighted mercury contamination in plants and animals in Ria de Aveiro (Portugal).

The last decades have observed a rapid increase in pharmaceutical loads entering the sewage system. Especially areas where treated wastewater and biosolids are used in agricultural fields have shown a high amount of pharmaceutical ingredients in the soil (Xia et al., 2005). There are a large number of studies available that have reported the occurrence of pharmaceuticals in natural water and investigated their impact (Balakrishna et al., 2017; Deo, 2014; Menon et al., 2020).

Indiscriminate use of pesticides in agricultural fields has led to severe contamination and adverse effects on nontarget organisms. Pesticides in the waterways can run off from the agricultural fields and end up in the waterways. The high content of pesticides can lead to algal bloom and create a hypoxic environment or dead zone. The presence of high content of pesticide has been reported worldwide and is a major concern and threat to human health (Albuquerque et al., 2016; Anju et al., 2010; de Souza et al., 2020). An increasing amount of pesticides can also interact with other contaminants and cause synergistic effects (Singh et al., 2017).

As plastic has occupied an indispensable position in the modern world, the contamination due to small plastic particles, i.e., microplastic/nanoplastics (MP/NP), has been a global issue. Their wide presence in all kinds of the natural environment has resulted in flooding research on its occurrence and its damaging effect on organisms (Bucci et al., 2020; Foley et al., 2018). Ingestion of MP/NP by organisms from different trophic levels and their interaction with other water contaminants is a serious threat to human health (Bhagat et al., 2021; Egbeocha et al., 2018). MP/NP can adsorb other contaminants on its surface and transport them a long distance.

1.6 Ecotoxicological Biomarkers

Biomarkers are used as a predictive tools to reveal the ecotoxicity of contaminants or stressors. Over the decades, environmentalists and toxicologists have given importance to developing biomarkers with ecological values. Biomarkers used in ecotoxicological studies provide the health status of an aquatic system by revealing

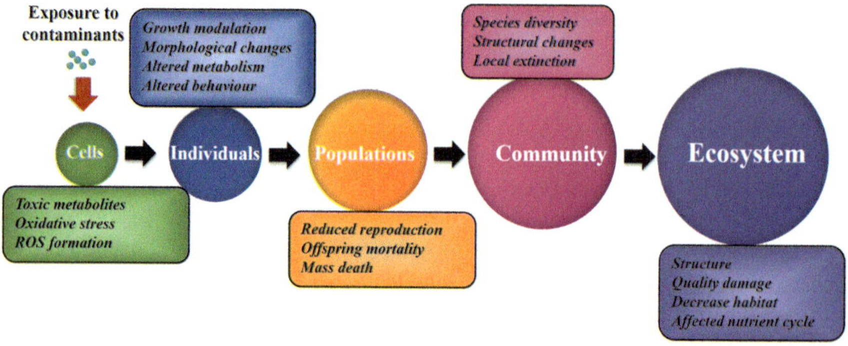

Fig. 1.3 Framework of ecotoxicological studies showing the effect of contaminants on different levels of organization

the presence of contaminants and their effects on organisms. Biomarkers can be divided into two major categories—defense biomarkers and damage biomarkers. The most common defense biomarkers include ethoxyresorufin-O-deethylase (EROD) activity, metallothioneins, phase II (glutathione S-transferase GST, N-acetyltransferase NAT, sulfotransferase SULT) enzymes, antioxidant defense, and heat shock proteins. Lysosomal membrane stability, acetylcholinesterase (AChE) activity, DNA damage, and thiobarbituric acid reactive substances (TBARS) are examples of biomarkers of damage.

Ecotoxicological studies use a combination of a battery of bioassays for assessing the contamination status and potential toxic effects associated with contaminants. The use of multiple biomarkers in ecological risk assessment studies results in the generation of a large amount of data that is hard to interpret. Hence, the integrated biomarker response (IBR) index is often calculated to assess the overall effect of contaminants (Bhagat et al., 2016; Meng et al., 2012). Ecotoxicological studies use different endpoints to measure the adverse effect of contaminants on the different biological organizations (individual, population, or community) (Fig. 1.3). Ecotoxicological effects of contaminants are measured on not only an individual organism but the entire ecosystem as a whole. Growth, reproduction, survival, behavioral pattern, and biochemical markers are commonly used ecotoxicological endpoints.

1.7 Bioindicators Used

Organisms living in the water of the contaminant are under constant stress, and their reaction to the stressor is influenced by the fluctuation in the contaminant's levels. Anthropogenic activities have resulted in the spillage of harmful substances into the natural environment. In this context, the use of bioindicators provides an early warning signal of the changes in the environment. Bioindicators are organisms or parts of the organism that is used to depict the quality of the environment, either positive or negative. In recent years, the rise in anthropogenic stressors has led to extensive research in finding suitable bioindicators. Several species from different biological taxonomic groups have been used as a bioindicator to reveal the health of the environment. Bioindicator species help in detecting the changes in the environment and in predicting the degree of contamination.

Bioindicator species are selected based on the type of contamination and the environment. A good bioindicator often shares properties such as

- Sensitivity toward changes in the environment.
- Sufficient population density with wide distribution.
- Toxicologically and ecologically well-studied.
- Easy to raise and bred.

There are three major groups of bioindicators:

1. Microbial indicators.
2. Plant indicators.
3. Animal indicators.

(f) **Microbial Indicators**

Microbes play a fundamental role in nutrient and energy cycling and maintain ecosystem health. They are the most abundant organisms and are crucially important for the stability of the biosphere. Microbes are used as an early warning signal to gain a better understanding of ecosystem health. The presence of fecal coliform and bacteriophage shows sewage contamination due to human and animal waste. Recent studies have shown the use of the bacterium *Vogesella indigofera* as an indicator of heavy metal contamination (Joseph, 2002). Other bacteria such as *Vibrio harveyi* have been proposed as potential bioindicators of mutagenic contamination (Czyz et al., 2000).

(g) **Plant Indicators**

The prevalence of plant communities in the aquatic system provides valuable information about ecosystem health. For example, the abundance of algae in a lake or river is often associated with the level of organic contaminants (nitrogen, phosphorus). Plant species such as planktons, aquatic plants, and lichens have increasingly been used as a bioindicator to depict the ecosystem status. Planktons are a crucial source of food for larger trophic-level organisms. Their wide presence in the aquatic ecosystem is efficiently used in understanding the contamination status of a given aquatic system. The presence of certain compounds in the water can promote or hinder the growth of plankton. The predominance of charophytes and Cyanophyta in the water indicates eutrophication of the aquatic ecosystem (Thakur et al., 2013). The presence of *Synechococcus leopoliensis*, *Dunaliella*, liverworts, and mosses is used as an indicator of heavy metal contamination. *Wolffia globosa*, *Eloderdensa*, and *Pontederia cordata* have been used for cadmium, mercury, and lead contamination. Due to their immobile nature and long life, higher plants have been used as a bioindicator to depict the ecosystem status of a localized area. The effect of water contaminants on plants can be observed using morphological characteristics as well as biochemical parameters.

(h) **Animal Indicators**

The presence or absence of animals may indicate the ecosystem's health status. For example, the lack of plankton and food sources in an area can significantly reduce the population of animal species. Hundreds of studies have used mussels as a bioindicator in environmental monitoring for assessing water quality. Mussel watch programs have been conducted worldwide for measuring polybrominated diphenyl ethers (Ramu et al., 2007), organic contaminants (Choi et al., 2010), heavy metals (Farrington et al., 1983), pesticides (Sericano et al., 1990), and microbial parasites (Kim & Powell, 2007). Over the past few decades, fish have been used efficiently to monitor the aquatic ecosystem's health. Fish are found in different habitats and have a long life cycle. They also occupy a wide range of trophic levels, which makes them important bioindicator organisms. They are sensitive to contaminants and respond significantly to changes in the environment. Fish communities have been used efficiently for heavy metal contamination (Birungi et al., 2007; Łuczyńska et al., 2018), toxic contamination (Weeks et al., 2018), and habitat degradation (Eugenia et al., 2019).

1.8 Risk Assessment

Risk assessment involves identification, analysis, and assessing the risk associated with a particular activity. Ecotoxicological risk assessment (ERA) is a relatively new field of research that studies the toxicity of contaminants and estimates the risk associated with their adverse effects. The main aim of an ERA is to evaluate possible adverse effects that may occur as a result of exposure due to stressors.

ERA involves three major steps—pre-risk assessment, risk assessment, and post-risk assessment (Fig. 1.4). Pre-risk assessment phase in ecotoxicological studies is the problem formulation phase which includes the identification of a possible problem that can occur due to a chemical or contaminant. In this phase, a background study is conducted to create baseline data for environmental quality. The second phase is the exposure and risk assessment phase, where a detailed study is conducted to investigate the bioavailability and effect of contaminants in model species. The mechanism underlying the toxic effects and its trophic transfer to higher-level organisms is studied. Various mathematic models and QSAR are used to predict ecotoxicity and its possible effect on human health. Laboratory ecotoxicity tests are conducted to determine the efficiency of effluent treatment plants. The nature of such tests can differ based on the type of effluents from industries such as mining, agrochemicals, plastic industries, and nuclear power plants. The last phase is the risk characterization phase. In this stage, results from toxicological studies are evaluated to estimate the potential adverse effects. The main objective of ERA is to balance risk and benefits and to set regulatory controls for industrial and

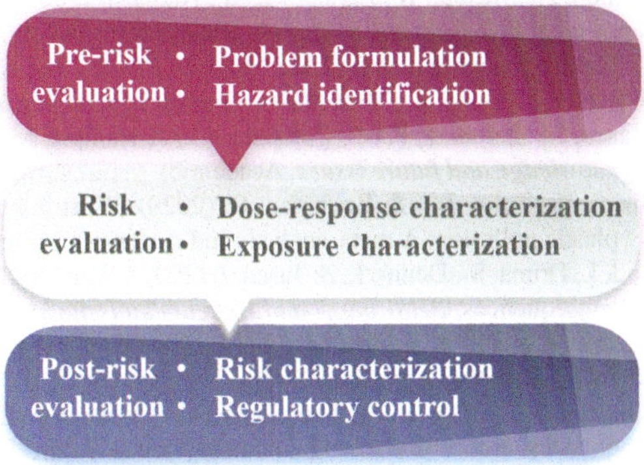

Fig. 1.4 A simplified methodology of ecotoxicological risk assessment studies

commercial activities. The ERA methodology adopted must be diverse to consider different types of risk and outcomes.

Overall, the concept of ecotoxicology and risk assessment is required for sustaining and enhancing the quality of human life. ERA helps in predicting and estimating the adverse effects of contaminants, but due to unforeseen stressors and its unpredictable effects, an ERA study alone is not sufficient. Hence, ecoepidemiology is often integrated, along with ERA, to identify and determine the causes of an impaired ecosystem.

References

Albuquerque, A., Ribeiro, J., Kummrow, F., Nogueira, A., Montagner, C., Umbuzeiro, G., et al. (2016). Pesticides in Brazilian freshwaters: A critical review. *Environmental Science: Processes & Impacts Journal, 18*, 779–787.

Ali, H., Khan, E., & Ilahi, I. (2019). Environmental chemistry and ecotoxicology of hazardous heavy metals: environmental persistence, toxicity, and bioaccumulation. *Journal of Chemistry, 2019*, 1.

Anju, A., Ravi, S. P., & Bechan, S. (2010). Water pollution with special reference to pesticide contamination in India. *Journal of Water Resource and Protection, 2010*.

Balakrishna, K., Rath, A., Praveenkumarreddy, Y., Guruge, K. S., & Subedi, B. (2017). A review of the occurrence of pharmaceuticals and personal care products in Indian water bodies. *Ecotoxicology and Environmental Safety, 137*, 113–120.

Baronti, C., Curini, R., D'Ascenzo, G., Di Corcia, A., Gentili, A., Samperi, R., et al. (2000). Monitoring natural and synthetic estrogens at activated sludge sewage treatment plants and in a receiving river water. *Environmental Science & Technology, 34*, 5059–5066.

Bhagat, J., Ingole, B., & Singh, N. (2016). Glutathione S-transferase, catalase, superoxide dismutase, glutathione peroxidase, and lipid peroxidation as biomarkers of oxidative stress in snails: A review. *ISJ-Invertebrate Survival Journal, 13*, 336–349.

Bhagat, J., Nishimura, N., & Shimada, Y. (2021). Toxicological interactions of microplastics/nanoplastics and environmental contaminants: Current knowledge and future perspectives. *Journal of Hazardous Materials, 405*, 123913.

Singh, J., Yadav, P., Pal, A. K., & Mishra, V. (2020). *Water pollutants: Origin and status. Sensors in water pollutants monitoring: Role of material* (pp. 5–20). Springer.

Thakur, R., Jindal, R., Singh, U. B., & Ahluwalia, A. (2013). Plankton diversity and water quality assessment of three freshwater lakes of Mandi (Himachal Pradesh, India) with special reference to planktonic indicators. *Environmental Monitoring and Assessment, 185*, 8355–8373.

UNEP. (2016). United Nations Environment Programme. A snapshot of the world's water quality: Towards a global assessment.

Valavanidis, A., & Vlachogianni, T. (2010). *University of Athens University Campus Zografou. Metal pollution in ecosystems. Ecotoxicology studies and risk assessment in the marine environment* (p. 15784). University of Athens.

Walker, C., et al. (1996). *Principles of ecotoxicology*. Taylor and Francis.

Weeks, B., Huggett, R., Warinner, J., & Mathews, E. (2018). *Macrophage responses of estuarine fish as bioindicators of toxic contamination. Biomarkers of environmental contamination* (pp. 193–201). CRC Press.

Xia, K., Bhandari, A., Das, K., & Pillar, G. (2005). Occurrence and fate of pharmaceuticals and personal care products (PPCPs) in biosolids. *Journal of Environmental Quality, 34*, 91–104.

Zafarzadeh, A., & Mehdinejad, M. (2015). Acecumulation of heavy metals in agricultural soil irrigated by sewage sludge and industrial effluent (case study: Agh Ghallah industrial estate). *Journal of Mazandaran University of Medical Sciences, 24*, 217–226.

Chapter 2
Nutrient Contamination in Marine Environment

Muhammed Nayeem Mullungal, Sajna Peediyakkathodi, Shazia Bibi, C. S. Ratheesh Kumar, and Mohammed Abu-Dieyeh

2.1 Introduction

Water is one of the most vital natural resources that sustains life, the foundation of ecology, and the key to primary production. Although water covers a majority of the Earth, freshwater makes up a less 2.5% of the total water on Earth from sources as diverse as rivers, lakes, groundwater, glaciers, and polar ice (Garcia Torres et al., 2022; Ghafari et al., 2008; Khatri & Tyagi, 2015). According to Singh, (2013) only a 0.6% of this freshwater can be used to make drinking water. The Clean Water Act, 1972, passed in the United States, led to many improvements in water bodies by curbing the toxic and organic pollutants released into the water (Andreen, 2003). However, more efforts are required to significantly reduce the major contaminants such as phosphorous and nitrogen in rivers, lakes, and coastal lines.

Aquatic species need nutrients to survive; however, too much nutrient input into water bodies might have an adverse effect on their intended usage (Bricker et al., 2007; Freeman et al., 2009; Rangel et al., 2016). Both natural and artificial factors cause nutrient contamination in water bodies. A number of factors involving source output and location, bioavailability, speciation and concentration, and seasonal changes influence nutrient enrichment to receiving water bodies (Withers & Jarvie, 2008). Since nitrogen (N) and phosphorus (P) are two significant nutrients, different species of these elements can be found, and when they move between the air, water,

M. N. Mullungal (✉) · S. Bibi · M. Abu-Dieyeh
Environmental Science Program, Department of Biological and Environmental Sciences, College of Arts and Sciences, Qatar University, Doha, Qatar
e-mail: mmullungal@qu.edu.qa

S. Peediyakkathodi · C. S. Ratheesh Kumar
School of Environmental Studies, Cochin University of Science and Technology, Kochi, Kerala, India

H. Alshemmari et al. (eds.), *Contaminated Land and Water*, Emerging Contaminants and Associated Treatment Technologies, https://doi.org/10.1007/978-3-031-65129-8_2

and soil, their forms can also alter. Even though N and P are essential for living organisms' normal growth and development, their overabundance would cause potentially harmful consequences. All species require the elements P and N, since they relate intricately to almost all ecological processes and serve as limiting elements for all trophic levels in the food chain (de Paula Filho et al., 2019; Valiela, 2009). Eutrophication and algal blooms can result from excess inputs of N and P into aquatic ecosystems by impacting algal biomass and community structure (Andersen et al., 2020; Wurtsbaugh et al., 2019). The chapter deals with nutrient contamination in water bodies, emphasizing their sources, biogeochemical cycling, consequences, and remedial measures.

2.2 Nutrients in the Aquatic Environment

It is established that nitrogen constitutes 78% of the atmosphere (Erisman et al., 2008) and is present in the air we breathe. It further spreads to the terrestrial ecosystem through nitrogen cycling and acid precipitation (Du et al., 2008). The biogeochemical cycle enhances nitrogen enrichment in the Earth and exhibits various forms such as NH_3, N_2, N_2O, NO, N_2O_3, NO_2, and N_2O_5 (Burghate & Ingole, 2014). N is a macroelement essential for the normal growth and development of the living system and constitutes biomolecules and genetic molecules (DNA, RNA), amino acids, vitamins, enzymes, etc. (Crews & Peoples, 2004). Nitrate (NO_3^-) and nitrite (NO_2^-) are the primary inorganic form of N in freshwater bodies, although nitrate has high solubility (Pitt et al., 1999), mobility (Martinez-Feria et al., 2019), and stability in water (Huan et al., 2020). Ammonia (NsH_3) and ammonium (NH_4^+) on the other hand are mostly toxic to aquatic life, relatively unstable though water-soluble (Weiss et al., 2008). The predominant presence of ammonia in water is in the form of NH_4^+ and undissociated NH_4OH with the latter being highly toxic to many organisms and especially so for fish (Trussell, 1972). The proportions of NH_4^+ to NH_4OH depend on the dissociation dynamics, which are governed by pH and temperature. Ammonia is strongly sorbed on particulate and colloidal particles, especially in alkaline lakes with high concentrations of humic dissolved organic matter.

Among other nutrients, phosphorus (P) has been recognized as one of the essential nutrients in the terrestrial and aquatic environments (Bauerfeind et al., 1990; Harrison et al., 1990) and is believed to be controlling aquatic primary productivity and influence biogeochemical cycle happening over a geological timescale (Ruttenberg, 2003; Scavia & Bricker, 2006). P, being a structural component of all organisms, forms the backbone of DNA and RNA and is crucial in transmitting chemical energy through ATP molecules (Mullungal et al., 2022). Phosphorus supply to the oceans exercises a critical control on primary productivity across global seas affecting organic carbon production and burial leading ultimately to the production of oxygen in time (Defforey & Paytan, 2018).

Phosphates (containing PO_4^{3-}) are the most prevalent P form present in natural waters, and they are only moderately soluble (USEPA, 1999) and are available in detergents, fertilizers, and pesticides. Its current abundance in the Earth's crust is approximately 0.12% (Aydin et al., 2009); however, almost all P on the Earth's crust is found in the form of minerals, including apatite (chlorapatite and fluorapatite), vivianite, wavellite, and phosphorite. Apatite {Ca (PO4)3[F, OH, or Cl]}, the largest reservoir of phosphate on Earth, is relatively insoluble in water (Bramha et al., 2014). Most of the P in the current natural environment is in particulate form and is not biologically available, limiting primary production. In marine waters, P is found in either dissolved or suspended forms; each has an inorganic form (orthophosphate, pyrophosphate, polyphosphate, and phosphate in minerals) and organic form (mono- and diester-, phosphonite). Organic and inorganic forms of phosphorus suffer a continuous transformation. Phytoplanktons absorb dissolved inorganic phosphorus (usually orthophosphate) which is then changed into organic phosphorus compounds (Aydin et al., 2009; Frankowski et al., 2002).

2.3 Potential Sources of Nutrient Contamination

The potential source of nutrient contamination is summed up in Fig. 2.1, and the details are shown below.

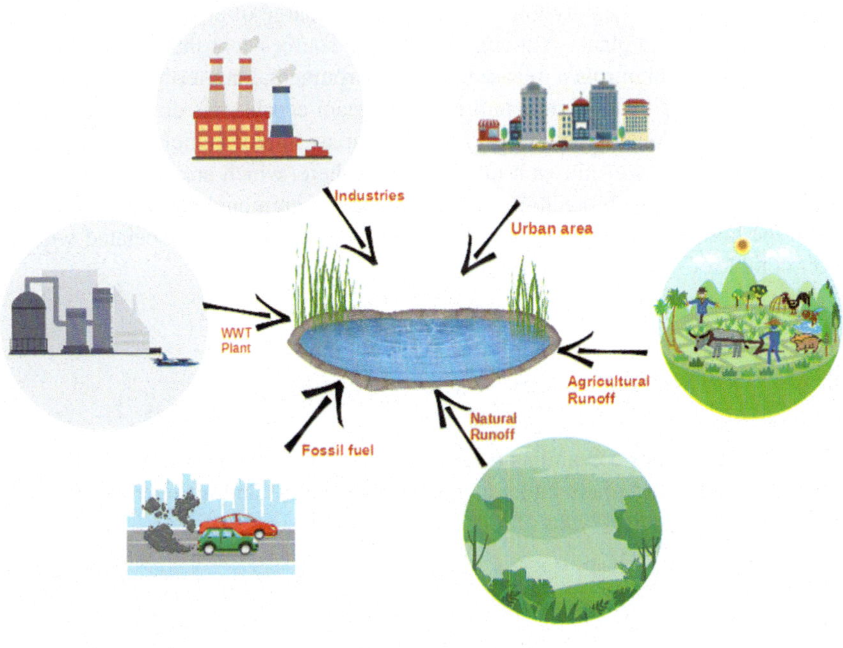

Fig. 2.1 The different sources of eutrophication

2.3.1 Agriculture

Chemicals (fertilizers and pesticides) applied in agroecosystems invariably enter the hydrological cycle, causing contamination of surface water as well as groundwater resources. Nitrates and pesticides (e.g., organophosphate) are common agrochemicals often identified as contaminants of groundwaters, whereas those of surface waters are sediments, pesticides, heavy metals, nitrogen, and phosphorus (Sharma, 1996). Animal manure, biocides, and chemical fertilizers contain P and N, obligated for crop growth. However, when plants do not fully utilize these nutrients, they can be lost from farm fields and harm the quality of air and downstream water. Plowing releases much N and P into the soil and the river, which also increases silt runoff and the nutrient content of the water. Nitrates are strongly connected with agricultural land and grasslands (Ferrier et al., 2001; Khatri & Tyagi, 2015), with the highest concentrations being in spring and in conjunction with high runoff events. The pesticide dispersal is impacted by topography, soil type, agricultural method, and crop type. In determining the runoff potential, each of these variables can outweigh the physicochemical properties of the constituent parts (Larson et al., 1995).

2.3.2 Industrial

Aquatic pollution is heavily influenced by the discharge of effluents from industry and water treatment plants. Although these point sources contribute only a tiny percentage of the total nitrogen released to the environment, long-term direct discharge to a water course could significantly affect stream ecosystems downstream of the discharge site (Rouse et al., 1999). In urban areas, motor vehicles and industrial exhaust emit oxides of nitrogen into the atmosphere, which are deposited into the aquatic system through precipitation. The release of ammonium, nitrate, and nitrite from the horticulture industry and phosphate is believed to be associated with the golf complex (Wayland et al., 2008).

2.3.3 Stormwater

Precipitation washes contaminants (such as N and P) from hard surfaces like rooftops, sidewalks, and roads into waterways. Phosphorus and nitrogen are the most common and impactful types of nutrient pollution in stormwater.

2.3.4 *Fossil Fuels*

Fossil fuels used in transportation, agriculture, electric power generation, and industry have increased the nitrogen content in the air. Since the combustion of fossil fuels emits oxidized nitrogen (NOx) into the atmosphere (Howarth, 2005), it contributes to the global budget for reactive nitrogen (Lucy et al., 2019). According to Ryther et al. (1972), electric power generation produces ~42% of N. In the United States, approximately 6.9 million metric tons of N annually emit into the environment through fossil fuel combustion (Howarth et al., 2002). The emitted N from fossil fuel is returned to the landscape through acid precipitation and dry deposition, putting up to nutrient contamination in water bodies (Howarth, 2005). Some researchers suggest that nitrogen emissions are reduced by encouraging less driving and more energy-efficient vehicles. Removing NOx from exhaust using catalytic converters and promoting electric power generation using fuel cells rather than traditional combustion would mitigate NOx emissions.

2.3.5 *Domestic Waste*

If not properly disposed off, yard and pet waste, certain soaps, and detergents can contribute to N and P, thereby leading to nutrient contamination (Fayiga & Nwoke, 2016). The hard surfaces and landscaping can enhance the runoff of N and P during wet weather.

2.3.6 *Natural Sources*

The Earth's volcanic gases also emit phosphorus as well as traces of methyl phosphine (CH_5P) (Morton et al., 2005). A baseflow can very well take the contamination from surface water to groundwater (Kwon et al., 2022).

2.4 Transport of Nutrients

Nitrogen and phosphorus are transported and transformed between the air, land, and water through the biogeochemical cycle. So, this session discusses the pathway of pollutants from the source to the water body. Nitrogen has inorganic and organic compounds in the biological system, environment, and industrial processes. Since both inorganic and organic nitrogen compounds can be found in biological systems, the environment, and industrial processes, the nitrogen cycle has a very important role in the ecosystem. In nature, large natural flows of the inorganic compounds of nitrogen from the atmosphere to the biological ecosystem through a physical and biological process produce ammonium compounds by unreactive molecular nitrogen reduction called the global nitrogen cycle (Fowler et al., 2013). The action of denitrifying organisms returns nitrogen gas to the atmosphere, which is enhanced by the microbes. Fixation, nitrification, and denitrification are three significant processes of the N cycle (Stein & Klotz, 2016), where nitrogen can be found as NH_3 (ammonia) and NH_4 (ammonium), all of which are stable compounds. The oxidation of nitrogen, in the nitrification stage of the nitrogen cycle (Fig. 2.2), converts NH_3 and NH_4 to NO_2^- (nitrite) and NO_3^- (nitrate) by the action of nitrifying bacteria, which represents the most stable forms of nitrogen (Takai, 2019).

Over the past 150 years, humans have significantly impacted the global P cycle, and this influence will likely continue for the foreseeable future. Since the anthropogenic modification began before scientific knowledge to quantify the P cycle, we could assume only P's "pre-anthropogenic" mass balance (Filippelli, 2002). The schematic diagram of the biogeochemical P cycle in the marine system is shown in Fig. 2.3.

Phosphorus is predominantly transported to the oceans by rivers, and up to 95% of P transported by rivers to estuaries and marine waters is particulate P (Föllmi, 1996). Understanding P transport from the land to the sea is vital for quantifying global P cycling (Föllmi, 1996; Ruttenberg & Berner, 1993) and while acting to mitigate the threats posed by eutrophication (Cloern, 2001). An intensively researched area regarding the P cycle has been the ecotones of coastal wetlands, estuaries and mangroves lying between land and ocean where the P cycle is much significance (Ashraf et al., 2006; Paludan & Morris, 1999; Rozan et al., 2002) (Prasad & Ramanathan, 2010).

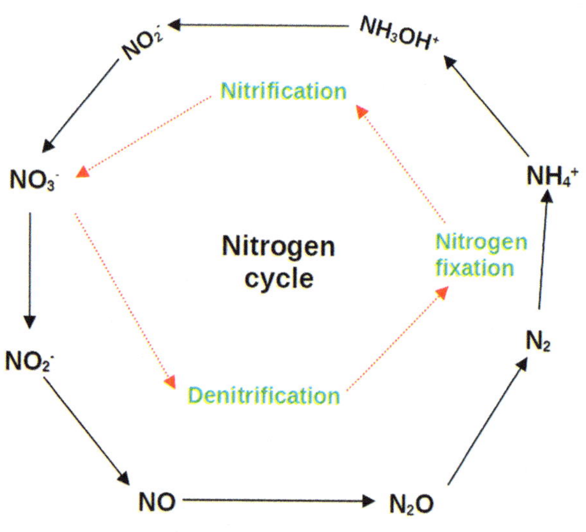

Fig. 2.2 Simplified scheme of the nitrogen cycle. (Modified Rosca et al., 2009)

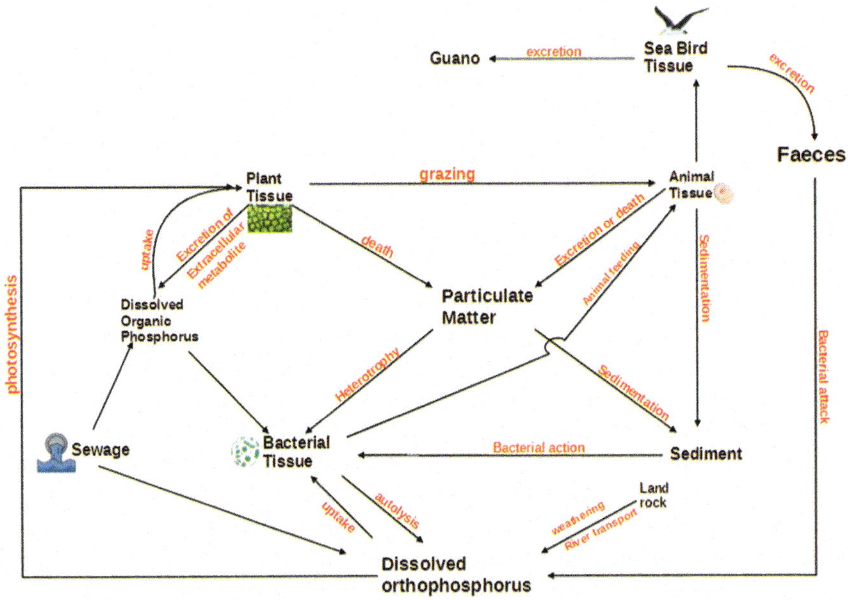

Fig. 2.3 Phosphorus cycle in marine system. (Modified Riley & Chester, 1971)

2.5 Eutrophication

The concentrations of nitrogen and phosphorus are known to play a vital role in determining the ecological status of aquatic systems (Jarvie et al., 1998), and their richness may lead to diverse problems such as an increase in the occurrence and extent of algal blooms, loss of oxygen, taste and odor problems, fish deaths, and loss of biodiversity. Nutrient enrichment destroys sustainability and the safe use of the aquatic ecosystem and impairs drinking water use (Carpenter et al., 1998). Eutrophication caused by over-enrichment with N and P is a widespread problem in rivers, lakes, reservoirs, and coastal regions, to which the global warming and alterations in precipitation patterns are adding further destructive effects (Sinha et al., 2017). Recent studies suggest that lake eutrophication in China was influenced by drastic changes in the season, climate anomalies, and variations in atmospheric-oceanic interactions (Qin et al., 2021).

OSPAR (1999) defines eutrophication as "Nutrients enrichment in the water that accelerates the growth of algae and higher forms of flora and fauna, resulting in an undesirable disturbance to the equilibrium of aquatic organisms and the quality of the water concerned, and thus refers to the severe complications resulting from anthropogenic addition of nutrients as described in the Common Procedure." The nutrient enrichment in water bodies has contributed to conditions like algal blooms, elevated bacterial levels, and increased biochemical oxygen demand (BOD). The reduction of water quality has altered species composition and the overall health of aquatic communities. Eutrophication sponsors a host of negative effects like increased levels in the biomass of aquatic vegetation, loss of commercial and sport fisheries, reduction in carbon and oxygen availability to food webs, habitat disruption, increase in the blooms of gelatinous zooplanktons (marine environment), increased toxins from bloom-forming algal species, loss of coral reefs, awful taste and odor along with depletions of DO, and economic loss to the tourism sector due to diminished aesthetic appeal of waterscapes (Smith & Schindler, 2009).

Oceans are inarguably a productive and valuable ecosystems on Earth not only because of their biological significance but also due to the economic output resulting from fishing and allied activities (Costanza et al., 1997). However, these ecosystems are frequently stressed because of conflicting human interests directly on the coast or their tributary rivers' catchments. Eutrophication in coastal and marine environments has been encouraging since the early 1980s (de Jonge & Elliot, 2001). The source of eutrophication and its impacts differs depending on the coastal systems (coastal marsh, estuaries, bays, or large coastal systems). A high influx of phytoplankton nutrients in the marine system, especially in enclosed seas like the Mediterranean Sea, may have led to harmful effects. In the marine environment, the nutrient enrichment stimulates the massive growth of phytoplankton, epiphytes (micro and/or macro), and free-floating opportunistic macroalgae (Valiela et al., 1997), which results in reduced light levels in the seagrasses beneath. Given that the compensation irradiance for seagrass growth is higher than that for most ephemeral macroalgae (Enríquez & Pantoja-Reyes, 2005), a light reduction of this sort and magnitude may result in negative seagrass growth (Olivé et al., 2009). Apart from this, coastal eutrophication can bring about anoxia and hypoxia, habitat destruction, reduction in the quality of water, loss of biodiversity, alteration of food web structure, and toxic and harmful algal blooms (Howarth, 2008; Rabalais et al., 2009). Also, coastal hypoxia contributes to ocean acidification, harming the calcifying organisms, such as mollusks and crustaceans. Hypoxic and anoxic water is linked with elevated carbon dioxide, which enhances acidification, accelerating the perturbation of ocean chemistry and influencing carbon dioxide emissions into the atmosphere (Howarth et al., 2011; Lucy et al., 2019). This nutrient imbalance alone might also influence the abundance, composition, virulence, and survival of pathogens that are already present in the aquatic ecosystems. Nutrient enrichment enhances the replication of the virus. Also, lesions in marine coral communities caused by infections by *Aspergillus* fungi grow at faster rates under high nitrate availability (Smith & Schindler, 2009).

2.6 Consequences

2.6.1 Harmful Algal Blooms (HABs)

Harmful algal blooms (HABs) are a natural phenomenon triggered by nutrient enrichment. However, the abundant supply of nutrients into coastal ecosystems has resulted in the excessive growth of filamentous macroalgae globally. As an impact, algae "blooms" mass drift is observed (Paalmea'b et al., 2004). These algal mats accumulate in very high biomasses (up to 400 kg wet weight m^{-2}) on the beaches and in estuaries, bays, and lagoons leading to massive ecological destruction due to the alterations happening to the nutrient dynamics of water columns and sediments (Briand, 1991). Furthermore, their decomposition results in anoxic conditions that are pervasive bringing in unstability to the entire shallow-water ecosystem, thereby impacting the higher trophic levels (Hansen & Kristensen, 1997; Paalme et al., 2002). Sometimes, the fish kills that result from these harmful algal blooms are simply a result of the pure numbers of algae involved when high algal respiration or high bacterial decomposition of the bloom leads to oxygen depletion. On other occasions, however, the algae have the capability to produce toxins which can kill fish or find their way through fish or shellfish to humans, resulting in various kinds of poisoning. Another type of harmful algal bloom (often termed noxious algal blooms) involves the algae damaging the fish gills, which can lead to severe losses in cage fish (Hodgkiss & Ho, 1997).

Smayda (1997) found that approximately 60–80 species of the 400 known phytoplankton produce toxins capable of harmful algal blooms. Algal toxin release could cause mortalities of fish, birds, and marine mammals as well as a human illness by consuming fisheries (James et al., 2010). In humans, algal bloom toxicity has mainly been reported from shellfish consumption (James et al., 2010). The most common and severe blooms are typically caused by cyanobacteria, the only known freshwater algae capable of producing toxins dangerous to human health. They can threaten human health and aquatic life. Toxins produced by cyanobacteria include neurotoxins like anatoxin-a and homoanatoxin-a, which are neuromuscular inhibiting agents. These toxins cause staggering, gasping, and muscle fasciculation in organisms. Shellfish contaminated by saxitoxin produced by dinoflagellate red tides can cause paralytic shellfish poisoning when contaminated shellfish are consumed (Codd, 2000). Threats further extend into the economic sector as well when damages of cyano-HABs include the loss of recreational revenue and increased drinking water treatment costs (Ferreira et al., 1999).

2.6.2 *Nutrient Toxicity*

The primary toxicity problem is the molecular form of ammonia (NH_3) which is harmful to aquatic life. A slight rise in pH may cause a tremendous increase in toxicity as the ammonium ion ($NH4^+$) is transformed into ammonia by the following equation (Zoungrana et al., 2020):

$$NH_4^+ + OH^- \rightarrow NH_3 + \mathrm{H_2O}$$

Dissolved oxygen and carbon dioxide concentrations, bicarbonate, and alkalinity may enhance ammonia toxicity at a given pH. The presence of excess ammonium and other nitrogen compounds at low salinity will jeopardize shrimp culture. During

total dissolved solids. Following breakpoint chlorination, dechlorination is frequently used to reduce the toxicity of the chlorine residual in the effluent. The advantage of this process is that ammonia nitrogen concentration can be reduced to near zero in the effluent.

2.9 Removal of Ammonium Ion

At present, there are several methods and technologies available for the ammonium removal which include biological method by converting ammonium into nitrate (nitrification) followed by nitrate conversion into nitrogen (denitrification) (Verstraete & Philips, 1998); the physicochemical approaches include air stripping (Gustin & Marinsek-Logar, 2011), nano-filtration (Marttinen et al., 2002), oxidation (Zhang et al., 2018), precipitation (Doyle and Parsons, 2002), adsorption (Huang et al., 2018), reverse osmosis, microwave radiation, and supercritical water oxidation. Because of its effectiveness, the selective ion exchange method can remove ammonium ions. In this treatment, ammonium passes through zeolite (Vocciante et al., 2018). Due to their high selectivity for ammonium ions, natural zeolites like mesolite, clinoptilotile, and sepiolite are used as ion exchangers (Huang et al. 2018).

Gaseous ammonia can be dissolved in water in varying degrees. Its molecular form is a gas that dissolves in water to varying degrees depending on the partial pressure of ammonia in the air adjacent to the water. Ammonia exits the water phase and enters the air when the partial pressure is reduced. Ammonia removal from wastewater can be accomplished by exposing small drops of water to a large amount of ammonia-free air the process called desorption. Ammonia must be in the molecular form (NH_3) rather than in the ammonium ion (NH_4^+) form in order to be stripped from wastewater. This is accomplished by raising the pH of the wastewater to 10 of 11, usually by the addition of lime. Addition of lime is often used for phosphate removal; it can serve a dual role. Again, nitrate nitrite, and organic N are not affected. The major issues associated with ammonia stripping are its inefficiency in cold weather, required shutdown during freezing conditions, and calcium carbonate scale formation in the air-stripping tower.

2.10 Removal of Phosphorus

Contemporary methods of P removal from water and wastewater involve a variety of techniques vis-à-vis physical and biological treatment, chemical precipitation ion exchange, and adsorption (Zahed et al., 2022).

Phosphate may be removed by using enhanced biological phosphate removal (EBPR) under alternating anaerobic and aerobic environments. In this environment, the microorganisms responsible for the bio-P removal process, known as polyphosphate-accumulating organisms (PAOs), are selected (Bassin et al., 2012; Van Loosdrecht et al., 1997). However, such conditions also favor the development of a particular group of microorganisms, called glycogen-accumulating organisms (GAOs), which compete with PAOs for the available organic substrate (Zeng et al., 2002). Decrease of phosphate removal performance in EBPR systems is often attributed to the presence of GAOs (Oehmen et al., 2006). Since the efficiency of P removal by microorganism depends on environmental factors, the removal of P from wastewater by biological treatments involve serious challenges.

Chemical precipitation produces large amount of sludge, while P removal and post-processes, such as separation, squeezing, and sludge disposal, are required (Jiang & Graham, 1998). Chemical precipitation of nutrients as struvite (magnesium ammonium phosphate) from various wastewater streams has been predominantly investigated for P recovery. Recovered struvite can be directly utilized as a slow-release fertilizer (Aydin et al., 2009; Costanza et al., 1997; de Jonge & Elliot, 2001; Wayland et al., 2008). Along with the addition of magnesium (Mg^{2+}) salt, struvite process may consume a significant amount of base chemicals due to the requirement of high pH conditions for struvite precipitation (Codd, 2000). Struvite process can provide up to 90% P recovery from wastewater with minimal N removal: theoretically, per gram of P removal as struvite removes 0.45 g of N (Aydin et al., 2009). In comparison with struvite, only a few studies reported efficient P recovery as calcium phosphate (CaP) from digested sludge centrate (Crews & Peoples, 2004; de Paula Filho et al., 2019). To the best of our knowledge, to date, no studies have compared P recovery efficiencies as struvite and calcium phosphate.

Among these, adsorption is considered to be the best option for the removal of P from aqueous solutions as there is maintenance as well as operating, highly efficient removal, minimal sludge production, lower costs, and a less intensive use of energy (Park et al., 2016; Lin et al., 2017). However, it is expensive to regenerate and dispose of the used adsorbents and can cause secondary environmental pollution.

References

Andersen, I. M., Williamson, T. J., González, M. J., & Vanni, M. J. (2020). Nitrate, ammonium, and phosphorus drive seasonal nutrient limitation of chlorophytes, cyanobacteria, and diatoms in a hyper-eutrophic reservoir. *Limnology and Oceanography, 65*(5), 962–978.

Andreen, W. L. (2003). Water quality today-has the clean water act been a success. *Alabama Law Review, 55*, 537.

Andersson, A., Hajdu, S., Haecky, P., Kuparinen, J., & Wikner, J. (1996). Succession and growth limitation of phytoplankton in the Gulf of Bothnia (Baltic Sea). *Marine Biology, 126*, 791–801.

Ashraf, M., Edwin, L., & Meenakumari, B. (2006). Studies on the seasonal changes of phosphorus in the marine environments off Cochin. *Environment International, 32*(2), 159–164. https://doi.org/10.1016/j.envint.2005.08.006

Aydin, I., Aydin, F., Saydut, A., & Hamamci, C. (2009). A sequential extraction to determine the distribution of phosphorus in the seawater and marine surface sediment. *Journal of Hazardous Materials, 168*(2–3), 664–669.

Bassin, J., Kleerebezem, R., Dezotti, M., & Van Loosdrecht, M. (2012). Simultaneous nitrogen and phosphate removal in aerobic granular sludge reactors operated at different temperatures. *Water Research, 46*(12), 3805–3816.

Bauerfeind, E., Hickel, W., Niermann, U., & Westernhagen, H. V. (1990). Phytoplankton biomass and potential nutrient limitation of phytoplankton development in the southeastern North Sea in spring 1985 and 1986. *Netherlands Journal of Sea Research, 25*(1–2), 131–142.

Bramha, S., Mohanty, A., Padhi, R., Panigrahi, S., & Satpathy, K. (2014). Phosphorus speciation in the marine sediment of Kalpakkam coast, southeast coast of India. *Environmental Monitoring and Assessment, 186*(10), 6003–6015.

Briand, X. (1991). Seaweed harvesting in Europe. In M. D. Guiry & G. Blunden (Eds.), *Seaweed resources in Europe: Uses and potential* (pp. 259–308). Wiley.

Bricker, S., Longstaff, B., Dennison, W., Jones, A., Boicourt, K., Wicks, C., & Woerner, J. (2007). Effects of nutrient enrichment in the Nation's estuaries: A decade of change (National Estuarine Eutrophication Update). *NOAA Coastal Ocean Program Decision Analysis Series, 26.*

Burghate, S., & Ingole, N. (2014). Biological denitrification—A review. *Journal of Environmental Science, Computer Science and Engineering & Technology, 3*(1), 009–028.

Carpenter, S. R., Caraco, N. F., Correll, D. L., Howarth, R. W., Sharpley, A. N., & Smith, V. H. (1998). Nonpoint pollution of surface waters with phosphorus and nitrogen. *Ecological Applications, 8*(3), 559–568.

Cloern, J. E. (2001). Our evolving conceptual model of the coastal eutrophication problem. *Marine Ecology Progress Series, 210*, 223–253.

Codd, G. A. (2000). Cyanobacterial toxins, the perception of water quality, and the prioritisation of eutrophication control. *Ecological Engineering, 16*(1), 51–60.

Costanza, R., d'Arge, R., De Groot, R., Farber, S., Grasso, M., Hannon, B., Limburg, K., Naeem, S., O'neill, R. V., & Paruelo, J. (1997). The value of the world's ecosystem services and natural capital. *Nature, 387*(6630), 253–260.

Crews, T. E., & Peoples, M. (2004). Legume versus fertilizer sources of nitrogen: Ecological tradeoffs and human needs. *Agriculture, Ecosystems & Environment, 102*(3), 279–297.

de Jonge, V., & Elliot, M. (2001). *Eutrophication of encyclopedia of ocean sciences 2* (pp. 852–870). Elsevier Ltd..

de Paula Filho, F. J., de Sá Sampaio, A. D., Menezes, J. M. C., da Costa, C. T. F., & Santiago, M. O. (2019). Land uses, nitrogen and phosphorus estimated fluxes in a Brazilian semi-arid watershed. *Journal of Arid Environments, 163*, 41–49.

Defforey, D., & Paytan, A. (2018). Phosphorus cycling in marine sediments: Advances and challenges. *Chemical Geology, 477*, 1–11.

Du, C., Zeng, G., Zhang, G., Tang, L., Li, X., Huang, D., Huang, L., & Jiang, Y. (2008). Input–output budgets for inorganic nitrogen under acid rain in a subtropical evergreen mixed forest in central-South China. *Water, Air, and Soil Pollution, 190*(1), 171–181.

Doyle, J. D., & Parsons, S. A. (2002). Struvite formation, control and recovery. *Water Research, 36*(16), 3925–3940. ISSN 0043-1354. https://doi.org/10.1016/S0043-1354(02)00126-4

EFSA. (2008). Nitrate in vegetables-scientific opinion of the panel on contaminants in the food chain. *EFSA Journal, 6*(6), 689.

Enríquez, S., & Pantoja-Reyes, N. I. (2005). Form-function analysis of the effect of canopy morphology on leaf self-shading in the seagrass Thalassia testudinum. *Oecologia, 145*(2), 234–242.

Erisman, J. W., Sutton, M. A., Galloway, J., Klimont, Z., & Winiwarter, W. (2008). How a century of ammonia synthesis changed the world. *Nature Geoscience, 1*(10), 636–639.

Fayiga, A. O., & Nwoke, O. (2016). Phosphate rock: Origin, importance, environmental impacts, and future roles. *Environmental Reviews, 24*(4), 403–415.

Ferreira, M., Franco, A., Catarino, L., Moreira, I., & Sousa, P. (1999). Environmental factors related to the establishment of algal mats in concrete irrigation channels. *Hydrobiologia, 415*, 163–168.

Ferrier, R. C., Edwards, A. C., Hirst, D., Littlewood, I. G., Watts, C. D., & Morris, R. (2001). Water quality of Scottish rivers: Spatial and temporal trends. *Science of the Total Environment, 265*(1), 327–342. https://doi.org/10.1016/S0048-9697(00)00674-4

Filippelli, G. M. (2002). The global phosphorus cycle. *Reviews in Mineralogy and Geochemistry, 48*(1), 391–425.

Föllmi, K. (1996). The phosphorus cycle, phosphogenesis and marine phosphate-rich deposits. *Earth-Science Reviews, 40*(1–2), 55–124.

Fowler, D., Coyle, M., Skiba, U., Sutton, M. A., Cape, J. N., Reis, S., Sheppard, L. J., Jenkins, A., Grizzetti, B., & Galloway, J. N. (2013). The global nitrogen cycle in the twenty-first century. *Philosophical Transactions of the Royal Society B: Biological Sciences, 368*(1621), 20130164.

Frankowski, L., Bolałek, J., & Szostek, A. (2002). Phosphorus in bottom sediments of pomeranian bay (southern Baltic—Poland). *Estuarine, Coastal and Shelf Science, 54*(6), 1027–1038.

Freeman, A. M., Lamon, E. C., III, & Stow, C. A. (2009). Nutrient criteria for lakes, ponds, and reservoirs: a Bayesian TREED model approach. *Ecological Modelling, 220*(5), 630–639.

Garcia Torres, E., Perez Morales, R., Gonzalez Zamora, A., Rios Sanchez, E., Olivas Calderon, E. H., de Alba Romero, J. J., & Calleros Rincon, E. Y. (2022). Consumption of water contaminated by nitrate and its deleterious effects on the human thyroid gland: a review and update. *International Journal of Environmental Health Research, 32*(5), 984–1001.

Ghafari, S., Hasan, M., & Aroua, M. K. (2008). Bio-electrochemical removal of nitrate from water and wastewater—a review. *Bioresource Technology, 99*(10), 3965–3974.

Gordon, H. R., Clark, D. K., Brown, J. W., Brown, O. B., Evans, R. H., & Broenkow, W. W. (1983). Phytoplankton pigment concentrations in the Middle Atlantic Bight: comparison of ship determinations and CZCS estimates. *Applied Optics, 22*, 20–36.

Guštin, S., & Marinšek-Logar, R. (2011). Effect of pH, temperature and air flow rate on the continuous ammonia stripping of the anaerobic digestion effluent. *Process Safety and Environmental Protection, 89*(1), 61–66. ISSN 0957-5820. https://doi.org/10.1016/j.psep.2010.11.001

Hansen, K., & Kristensen, E. (1997). Impact of macrofaunal recolonization on benthic metabolism and nutrient fluxes in a shallow marine sediment previously overgrown with macroalgal Mats. *Estuarine, Coastal and Shelf Science, 45*(5), 613–628. https://doi.org/10.1006/ecss.1996.0229

Harrison, P. J. H., Hu, M. H., Yang, Y. P., & Lu, X. (1990). Phosphate limitation in estuarine and coastal waters of China. *Journal of Experimental Marine Biology and Ecology, 140*, 79–87.

Hodgkiss, I. J., & Ho, K. C. (1997). Are changes in N:P ratios in coastal waters the key to increased red tide blooms? *Hydrobiologia, 352*, 141–147. https://doi.org/10.1023/A:1003046516964

Howarth, R. W. (2005). The development of policy approaches for reducing nitrogen pollution to coastal waters of the USA. *Science China Life Sciences, 48*, 791–806.

Howarth, R. W. (2008). Coastal nitrogen pollution: a review of sources and trends globally and regionally. *Harmful Algae, 8*(1), 14–20.

Howarth, R. W., Boyer, E. W., Pabich, W. J., & Galloway, J. N. (2002). Nitrogen use in the United States from 1961–2000 and potential future trends. *Ambio, 31*(2), 88–96. http://www.jstor.org/stable/4315220

Howarth, R., Chan, F., Conley, D. J., Garnier, J., Doney, S. C., Marino, R., & Billen, G. (2011). Coupled biogeochemical cycles: Eutrophication and hypoxia in temperate estuaries and coastal marine ecosystems. *Frontiers in Ecology and the Environment, 9*(1), 18–26. https://doi.org/10.1890/100008

Huan, H., Hu, L., Yang, Y., Jia, Y., Lian, X., Ma, X., Jiang, Y., & Xi, B. (2020). Groundwater nitrate pollution risk assessment of the groundwater source field based on the integrated numerical simulations in the unsaturated zone and saturated aquifer. *Environment International, 137*, 105532.

Huang, J., Kankanamge, N. R., Chow, C., Welsh, D. T., Li, T., & Teasdale, P. R. (2018). Removing ammonium from water and wastewater using cost-effective adsorbents: A review. *Journal of Environmental Sciences, 63*, 174–197. ISSN 1001-0742. https://doi.org/10.1016/j.jes.2017.09.009

Jaanus, A., Toming, K., Hällfors, S., Kaljurand, K., & Lips, I. (2009). Potential phytoplankton indicator species for monitoring Baltic coastal waters in the summer period. In *Eutrophication in coastal ecosystems* (pp. 157–168). Springer.

James, K., Carey, B., O'halloran, J., & Škrabáková, Z. (2010). Shellfish toxicity: Human health implications of marine algal toxins. *Epidemiology & Infection, 138*(7), 927–940.

Jarvie, H., Whitton, B., & Neal, C. (1998). Nitrogen and phosphorus in east coast British rivers: Speciation, sources and biological significance. *Science of the Total Environment, 210*, 79–109.
Jiang, J. Q., & Graham, N. J. D. (1998). Pre-polymerised inorganic coagulants and phosphorus removal by coagulation: a review. *Water SA, 24*(3), 237–244.
Kahru, M., & Leppanen, J.-M. (1993). Cyanobacterial blooms cause heating of the sea surface. *Marine Ecology Progress Series, 101*, 1–7. https://doi.org/10.3354/meps101001
Khatri, N., & Tyagi, S. (2015). Influences of natural and anthropogenic factors on surface and groundwater quality in rural and urban areas. *Frontiers in life Science, 8*(1), 23–39.
Kwon, H., Koh, D. C., Cho, B. W., & Jung. Y. Y. (2022). Nutrient dynamics in stream water and groundwater in riparian zones of a mesoscale agricultural catchment with intense seasonal pumping. *Agricultural Water Management, 261*, 107336. https://doi.org/10.1016/j.agwat.2021.107336
Larson, S. J., Capel, P. D., Goolsby, D. A., Zaugg, S. D., & Sandstrom, M. W. (1995). Relations between pesticide use and riverine flux in the Mississippi River basin. *Chemosphere, 31*(5), 3305–3321. https://doi.org/10.1016/0045-6535(95)00176-9
Liu, X y., & Hong, Y. (2021). Microalgae-Based Wastewater Treatment and Recovery with Biomass and Value-Added Products: a Brief Review. *Curr Pollution Rep, 7*, 227–245. https://doi.org/10.1007/s40726-021-00184-6
Lin, J., Zhan, Y., Wang, H., Chu, M., Wang, C., He, Y. et al. (2017). Effect of calcium ion on phosphate adsorption onto hydrous zirconium oxide. *Chemical Engineering Journal, 309*, 118–129.
Lindahl, O., & Kollberg, S. (2009). Can the EU agri-environmental aid program be extended into the coastal zone to combat eutrophication? *Hydrobiologia, 629*, 59–64. https://doi.org/10.1007/s10750-009-9771-3
Lucy, N., Johnny, M. G., III, Daniel, M., & Robert, T. (2019). Nitrogen and phosphorus eutrophication in marine ecosystems. In F. Houma Bachari (Ed.), *Monitoring of marine pollution*. (pp. Ch. 5. IntechOpen. https://doi.org/10.5772/intechopen.81869
Martinez-Feria, R., Nichols, V., Basso, B., & Archontoulis, S. (2019). Can multi-strategy management stabilize nitrate leaching under increasing rainfall? *Environmental Research Letters, 14*(12), 124079. https://doi.org/10.1088/1748-9326/ab5ca8
Marttinen, S. K., Kettunen, R. H., Sormunen, K. M., Soimasuo, R. M., & Rintala, J. A. (2002). Screening of physical–chemical methods for removal of organic material, nitrogen and toxicity from low strength landfill leachates. *Chemosphere, 46*(6), 851–858. ISSN 0045-6535. https://doi.org/10.1016/S0045-6535(01)00150-3
Millie, D. F., Schofield, O. M., Dionigi, C. P., & Johnson, P. B. (1995). Assessing noxious phytoplankton in aquaculture systems using bio-optical methodologies: a review. *Journal of the World Aquaculture Society, 26*, 329–345.
Morton, S. C., Zhang, Y., & Edwards, M. A. (2005). Implications of nutrient release from iron metal for microbial regrowth in water distribution systems. *Water Research, 39*(13), 2883–2892. https://doi.org/10.1016/j.watres.2005.05.024
Mullungal, M. N., Thalayappil, S., Peediyakkathodi, S., Salas, P. M., Sujatha, C. H., & Kumar, C. S. R. (2022). Vertical distribution of phosphorous fractions and bioavailability of the nutrient in the southern Indian Ocean. *International Journal of Environmental Research, 16*(5), 77. https://doi.org/10.1007/s41742-022-00448-z
National Air Quality and Emissions Trends Report, 1998, EPA-454/R-00-003, U.S. Environmental Protection Agency, Office of Air Quality Planning and Standards, Research Triangle Park, NC 27711, March 2000.
Nujić, M., & Habuda-Stanić, M. (2017). Nitrates and nitrites, metabolism and toxicity. *Food in Health and Disease, 6*, 63–73.
Oehmen, A., Saunders, A. M., Vives, M. T., Yuan, Z., & Keller, J. (2006). Competition between polyphosphate and glycogen accumulating organisms in enhanced biological phosphorus removal systems with acetate and propionate as carbon sources. *Journal of Biotechnology, 123*(1), 22–32. https://doi.org/10.1016/j.jbiotec.2005.10.009
Olivé, I., García-Sánchez, M. P., Brun, F. G., Vergara, J. J., & Pérez-Lloréns, J. L. (2009). Interactions of light and organic matter under contrasting resource simulated environments:

The importance of clonal traits in the seagrass Zostera noltii. *Hydrobiologia, 629*(1), 199–208. https://doi.org/10.1007/s10750-009-9773-1
OSPAR Commission. (1999). OSPAR strategy with regard to hazardous substances, Sintra (Portugal), 22–23 July1998, Annex 34.
Paalme, T., Kukk, H., Kotta, J., & Orav, H. (2002). 'In vitro' and 'in situ' decomposition of nuisance macroalgae Cladophora glomerata and Pilayella littoralis. In E. Orive, M. Elliott, & V. N. de Jonge (Eds.), *Nutrients and eutrophication in estuaries and coastal waters: Proceedings of the 31st symposium of the estuarine and coastal sciences association (ECSA), held in Bilbao, Spain, 3–7 July 2000* (pp. 469–476). Springer. https://doi.org/10.1007/978-94-017-2464-7_36
Paalme, T., Martina, G., Kottaa, J., Kukk, H., & Kaljurand, K. (2004). Distribution and dynamics of drifting macroalgal mats in Estonian coastal waters during. *Proceedings of the Estonian Academy of Sciences, Biology and Ecology*,
Paludan, C., & Morris, J. T. (1999). Distribution and speciation of phosphorus along a salinity gradient in intertidal marsh sediments. *Biogeochemistry, 45*(2), 197–221. https://doi.org/10.1007/BF01106781
Park, T., Ampunan, V., Lee, S., & Chung, E. (2016). Chemical behavior of different species of phosphorus in coagulation. *Chemosphere, 144*, 2264–2269. ISSN 0045-6535. https://doi.org/10.1016/j.chemosphere.2015.10.131
Pitt, R., Clark, S., & Field, R. (1999). Groundwater contamination potential from stormwater infiltration practices. *Urban Water, 1*(3), 217–236. https://doi.org/10.1016/S1462-0758(99)00014-X
Prasad, K. B. K., Dittmar, T., & Ramanathan, A. L. (2010). Organic matter and mangrove productivity. In: A. L. Ramanathan, P. Bhattacharya, T. Dittmar, M. B. K. Prasad, & B. R. Neupane (Eds.), *Management and sustainable development of coastal zone environments* (pp. 175–193). Springer-Verlag, Berlin. https://doi.org/10.1007/978-90-481-3068-9_12
Qin, B., Deng, J., Shi, K., Wang, J., Brookes, J., Zhou, J., Zhang, Y., Zhu, G., Paerl, H. W., & Wu, L. (2021). Extreme climate anomalies enhancing cyanobacterial blooms in eutrophic Lake Taihu, China. *Water Resources Research, 57*(7), e2020WR029371. https://doi.org/10.1029/2020WR029371
Rabalais, N. N., Turner, R. E., Díaz, R. J., & Justić, D. (2009). Global change and eutrophication of coastal waters. *ICES Journal of Marine Science, 66*(7), 1528–1537. https://doi.org/10.1093/icesjms/fsp047
Rangabhashiyam, S., Anu, N., Giri Nandagopal, M. S., & Selvaraju, N. (2014). Relevance of isotherm models in biosorption of pollutants by agricultural byproducts. *Journal of Environmental Chemical Engineering, 2*(1), 398–414. https://doi.org/10.1016/j.jece.2014.01.014
Rangel, C. M. A., Neto, J. A. B., de Oliveira, P. S., Lima, L. G., & da Fonseca, E. M. (2016). Assessment of human effects through phosphorus partitioning in sediments from two catchment basins around Guanabara Bay, SE Brazil. *Environmental Earth Sciences, 75*(17), 1215. https://doi.org/10.1007/s12665-016-6015-5
Riley, J. P., & Chester, R. (1971). *Introduction to marine chemistry*. Academic Press, London and New York.
Rosca, V., Duca, M., de Groot, M. T., & Koper, M. T. (2009). Nitrogen cycle electrocatalysis. *Chemical Reviews, 109*(6), 2209–2244. https://doi.org/10.1021/cr8003696. PMID: 19438198.
Rouse, J. D., Bishop, C. A., & Struger, J. (1999). Nitrogen pollution: An assessment of its threat to amphibian survival. *Environmental Health Perspectives, 107*(10), 799–803. https://doi.org/10.1289/ehp.99107799
Rozan, T. F., Taillefert, M., Trouwborst, R. E., Glazer, B. T., Ma, S., Herszage, J., Valdes, L. M., Price, K. S., & Luther, G. W., III. (2002). Iron-sulfur-phosphorus cycling in the sediments of a shallow coastal bay: Implications for sediment nutrient release and benthic macroalgal blooms. *Limnology and Oceanography, 47*(5), 1346–1354. https://doi.org/10.4319/lo.2002.47.5.1346
Ruttenberg, K. (2003). The global phosphorus cycle. *Treatise on geochemistry, 8*, 682.
Ruttenberg, K. C., & Berner, R. A. (1993). Authigenic apatite formation and burial in sediments from non-upwelling, continental margin environments. *Geochimica et Cosmochimica Acta, 57*(5), 991–1007. https://doi.org/10.1016/0016-7037(93)90035-U

Ryther, J. H., Dunstan, W. M., Tenore, K. R., & Huguenin, J. E. (1972). Controlled eutrophication—increasing food production from the sea by recycling human wastes. *BioScience, 22*(3), 144–152. https://doi.org/10.2307/1296076

Scavia, D., & Bricker, S. B. (2006). Coastal eutrophication assessment in the United States. *Biogeochemistry, 79*(1/2), 187–208. http://www.jstor.org/stable/20519826

Sharma, M. (1996). Impact of agriculture on nutrient contamination of water resources. In *Water-quality hydrology* (pp. 57–79). Springer.

Shearer, L. A., Goldsmith, J. R., Young, C., Kearns, O. A., & Tamplin, B. R. (1972). Methemoglobin levels in infants in an area with high nitrate water supply. *American Journal of Public Health, 62*(9), 1174–1180. https://doi.org/10.2105/ajph.62.9.1174

Singh, A. (2013). Groundwater modelling for the assessment of water management alternatives. *Journal of Hydrology, 481*, 220–229.

Sinha, E., Michalak, A. M., & Balaji, V. (2017). Eutrophication will increase during the 21st century as a result of precipitation changes. *Science, 357*(6349), 405–408. https://doi.org/10.1126/science.aan2409

Smayda, T. (1997). What is a bloom? A commentary. *Limnology and Oceanography, 42*, 1132–1136. https://doi.org/10.4319/lo.1997.42.5_part_2.1132

Smith, V. H., & Schindler, D. W. (2009). Eutrophication science: Where do we go from here? *Trends in Ecology & Evolution, 24*(4), 201–207. https://doi.org/10.1016/j.tree.2008.11.009

Stein, L. Y., & Klotz, M. G. (2016). The nitrogen cycle. *Current Biology, 26*(3), R94–R98. https://doi.org/10.1016/j.cub.2015.12.021

Takai, K. (2019). The nitrogen cycle: A large, fast, and mystifying cycle. *Microbes and Environments, 34*(3), 223–225. https://doi.org/10.1264/jsme2.ME3403rh

Trussell, R. P. (1972). The percent un-ionized ammonia in aqueous ammonia solutions at different pH levels and temperatures. *Journal of the Fisheries Research Board of Canada, 29*(10), 1505–1507. https://doi.org/10.1139/f72-236

USEPA (United States Environmental Protection Agency). (1999). National recommended water quality criteria–correction. EPA-822-Z-99-001. Office of Water. Washington, District of Columbia.

Valencia-Castañeda, G., Frías-Espericueta, M. G., Vanegas-Pérez, R. C., Pérez-Ramírez, J. A., Chávez-Sánchez, M. C., & Páez-Osuna, F. (2018). Acute toxicity of ammonia, nitrite and nitrate to shrimp Litopenaeus vannamei Postlarvae in low-salinity water. *Bulletin of Environmental Contamination and Toxicology, 101*(2), 229–234. https://doi.org/10.1007/s00128-018-2355-z

Valiela, I. (2009). *Global coastal change*. Wiley.

Valiela, I., McClelland, J., Hauxwell, J., Behr, P. J., Hersh, D., & Foreman, K. (1997). Macroalgal blooms in shallow estuaries: Controls and ecophysiological and ecosystem consequences. *Limnology and Oceanography, 42*(5part2), 1105–1118. https://doi.org/10.4319/lo.1997.42.5_part_2.1105

Van Loosdrecht, M., Hooijmans, C., Brdjanovic, D., & Heijnen, J. (1997). Biological phosphate removal processes. *Applied Microbiology and Biotechnology, 48*(3), 289–296.

Verstraete, W., & Philips, S. (1998). Nitrification-denitrification processes and technologies in new contexts. *Environmental Pollution, 102*(1, Supplement 1), 717–726. https://doi.org/10.1016/S0269-7491(98)80104-8

Vocciante, M., D'Auris, A. D. F., Finocchi, A., Tagliabue, M., Bellettato, M., Ferrucci, A., Reverberi, A. P., Ferro, S. (2018). Adsorption of ammonium on clinoptilolite in presence of competing cations: Investigation on groundwater remediation. *Journal of Cleaner Production, 198*, 480–487. ISSN 0959-6526. https://doi.org/10.1016/j.jclepro.2018.07.025

Wayland, D., Megson, D., Mudge, S., Icely, J., & Newton, A. (2008). Identifying the source of nutrient contamination in a lagoon system. *Environmental Forensics, 9*(2–3), 231–239.

Weiss, P. T., LeFevre, G., & Gulliver, J. S. (2008). Contamination of soil and groundwater due to stormwater infiltration practices, a literature review.

Withers, P. J. A., & Jarvie, H. P. (2008). Delivery and cycling of phosphorus in rivers: A review. *Science of the Total Environment, 400*(1), 379–395. https://doi.org/10.1016/j.scitotenv.2008.08.002

Wurtsbaugh, W. A., Paerl, H. W., & Dodds, W. K. (2019). Nutrients, eutrophication and harmful algal blooms along the freshwater to marine continuum. *WIREs Water, 6*(5), e1373. https://doi.org/10.1002/wat2.1373

Zahed, M. A., Salehi, S., Tabari, Y., Farraji, H., Ataei-Kachooei, S., Zinatizadeh, A. A., Kamali, N., & Mahjouri, M. (2022). Phosphorus removal and recovery: State of the science and challenges. *Environmental Science and Pollution Research, 29*(39), 58561–58589. https://doi.org/10.1007/s11356-022-21637-5

Zeng, R., Yuan, Z., Van Loosdrecht, M. C., & Keller, J. (2002). Proposed modifications to metabolic model for glycogen-accumulating organisms under anaerobic conditions. *Biotechnology and Bioengineering, 80*(3), 277–279. https://doi.org/10.1002/bit.10370

Zhang, L. M., Duff, A. M., & Smith, C. J. (2018). Community and functional shifts in ammonia oxidizers across terrestrial and marine (soil/sediment) boundaries in two coastal bay ecosystems. *Environ Microbiol 20*(8), 2834–2853. https://doi.org/10.1111/1462-2920.14238

Zhang, W., Tooker, N. B., & Mueller, A. V. (2020). Enabling wastewater treatment process automation: leveraging innovations in real-time sensing, data analysis, and online controls. *Environmental Science: Water Research & Technology, 6*(11), 2973–2992. https://doi.org/10.1039/D0EW00394H

Zoungrana, A., Zengin, İ. H., Türk, O. K., & Çakmakcı, M. (2020). Ammonium recovery performance of membrane distillation from synthetic solution. *Chemcial Papers, 74*, 1903–1915.

Chapter 3
Contaminant Types and Nature of Sources in General Aspects

T. P. Maneesh

3.1 Introduction

In recent decades, the fast growing of the world population and the pursuance of economic condition have generated a solid elaboration in industrial and agricultural production. Due to the overuse of chemical fertilizers and industrialization, our land and water become contaminated with various types of contaminants. The contaminants are those substances which adversely alter the environment; which are toxic to the human, animal, and plant health; and which are present in the environment above the permissible limits of concentration. The contaminants are formed by anthropogenic activities which include solid, liquid, and gaseous matter for interim economic benefits at the cost of enduring ecological benefits for humanity. An environmental contaminant may be chemical in nature; also, it may also be a biological (pathogens) or physical (energy) agent (Vallero, 2010).

The related increase in waste generation and energy consumption has massively increased the burden on the natural environment and has led to changes in the composition of the air, land, and water. This has led to disruption of natural ecosystems and a decline of environmental quality. The growing density of population has made the society increasingly insecure to the natural change in the environment.

Contaminated land is any area that has been polluted by its past industrial use or by the disposal of wastes. Contamination is an inevitable effect of industrialization and of waste generation, as it involves the introduction of materials not naturally present in the environment at a given location.

Contamination of water sources by substances which make the water unusable for drinking, cooking, cleaning, swimming, and other activities results to water pollution. The major contaminants include chemicals, trash, bacteria, and parasites.

T. P. Maneesh (✉)
Central Water Commission, Coimbatore, India

H. Alshemmari et al. (eds.), *Contaminated Land and Water*, Emerging Contaminants and Associated Treatment Technologies, https://doi.org/10.1007/978-3-031-65129-8_3

Sooner or later, all types of pollution finally reach to water. Air pollution drops on lakes and oceans. Land pollution can reach into an underground stream, then to a river, and ultimately to the ocean.

The sources of contaminants may be point source or nonpoint source. The common sources of contaminants are fossil fuels, industries and industrial accidents, oil spills, mining, ammunitions and agents of war, etc. The contaminants may be natural or anthropogenic in nature. The common environmental contaminants are polyaromatic hydrocarbons, heavy metals, pesticides, organic and inorganic solvents, etc. Accidental release of these contaminants in the environment leads to evolution of new diseases to human health and mass death of population. The present chapter deals with the various types of contaminants and nature of sources in land and water.

3.2 Types of Contaminants

The types of contaminants based on occurrence, existence, persistence, and chemical composition (Savita et al., 2019) are depicted in Fig. 3.1.

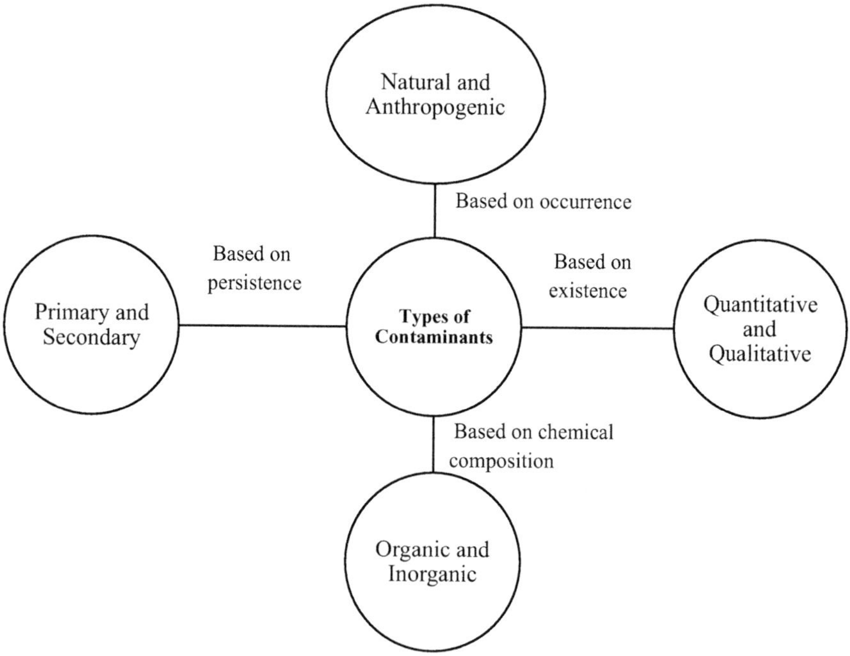

Fig. 3.1 The types of contaminants based on occurrence, existence, persistence, and chemical composition

The most common contaminants can be classified as natural and anthropogenic depending upon the occurrence. The substances found naturally in the environment are called natural contaminants, and the substances found in the environment due to human activities (Rhind, 2009) are called anthropogenic contaminants.

3.3 Land Contaminants

Land contamination is the deposition of solid/liquid wastes on land/underground that can contaminate the soil and groundwater or polluted by its past industrial use. Induction of materials not naturally present in the environment at a particular location leads to contamination which is an inevitable outcome of waste generation and industrialization (Doyle, 2021). Generally, land contamination can be classified as natural and anthropogenic (man-made).

3.4 Natural Contaminants

The natural accumulation of chemicals in rare case leads to contamination in soil/land. Also, the natural processes may have a control on human-released toxic substances (chemicals) into the soil, thereby decreasing/increasing the substance toxicity and/or the level of contamination of the soil/land due to the complex environment of soil. The main natural processes leading to land contaminations are natural accumulation, natural production, and leakage into subsurface. Also, the biological organisms (microbes) act on the human, bird, and animal excreta inside the soil to produce compost and slurry into the soil.

3.5 Anthropogenic Contaminants

Anthropogenic contamination is derived from various types of processes, such as industrial and accidental. Anthropogenic land contaminants may work in combination with natural processes to enhance the toxic contamination levels in land.

There are several reasons for land contamination which includes chemical waste dumping; mining activities; agricultural activities by the usage of pesticides and/or insecticides and fertilizers; accidental spills and leaks during storage, transport, or use of chemicals; construction activities including demolition; storage of waste in landfills and manufacturing processes that involve furnaces; or other processes resulting in the possible dispersion of contaminants in the environment.

3.6 Water Contaminants

Water contamination refers to the presence of substances which deteriorate the quality of fresh or marine water which includes water bodies such as lakes, streams, rivers, oceans, and groundwater. Water contamination can be natural, such as animal waste, or anthropogenic. The main categories of water contaminants are physical, chemical, biological, and radiological.

3.6.1 *Physical Contaminants*

The physical nature of water can change by the activity of physical contaminants. The natural disasters like landslides and floods deteriorate the water quality of rivers, lakes, and estuaries and alter the turbidity and other water quality properties (Knap & Rusyn, 2016). Human intervention like demolition of buildings near to water body can also deteriorate water quality as a result of waste falling into the water (Menon et al., 2021).

3.6.2 *Chemical Contaminants*

Chemical contaminants can be found as organic and inorganic substances in products commonly used by almost people in their day-to-day life. The common substances such as plastics, resins, pharmaceuticals, disinfectants, deodorants, detergents, petroleum products, road runoff, pesticides, and biocides are falling in this category of pollutants. It can also include naturally occurring chemicals like nitrogen and chlorine. There are two types of chemical contaminants, namely, organic and inorganic contaminants.

3.6.3 *Organic Contaminants*

Several organic compounds enter into water bodies as a consequence of anthropogenic activities, and they have diverse physical, chemical, and toxicological properties. Organic contaminants include mineral oil, petroleum products, phenols, pesticides, polychlorinated biphenyls (PCBs), surfactants, oil and petrol spills from roads and concreted areas, hormones, pesticides, herbicides, and fungicides deriving from agricultural and horticultural activities that are situated near the waterways. Organic contaminants are present in water as single molecules dissolved in water as well as suspended solids. The byproducts of industrial processes and petroleum production such as synthetic and volatile organic chemicals are also considered as organic contaminants.

Particularly, persistent organic pollutants (POPs) like organochlorine pesticides (OCPs), polybrominated diphenyl ethers (PBDEs), polyaromatic hydrocarbons (PAHs), polychlorinated naphthalenes (PCNs), polychlorinated biphenyls (PCBs), dechlorane plus (DP), etc. are a matter of concern (Akhtar et al., 2021) in recent decade while considering environmental contamination. The long-term impact of POP contamination adversely affects water bodies because of their nature such as persistent, toxic, ability of long-range transport, carcinogenic, lipophilic, and bioaccumulation.

Nowadays, contaminants of emerging concern (CECs) are commonly found in aquatic systems, which include pharmaceuticals, personal care products, endocrine-disrupting compounds, plasticizers, flame retardants, etc., and are of great concern to the environment. CECs are present in surface water (Nilsen et al., 2019), groundwater (Cabeza et al., 2012), drinking water (Yang et al., 2014), and wastewater treatment plant discharge (Samaras et al., 2013). The major source of CECs into the environment is municipal wastewater as well as point and nonpoint sources, industries, and wastewater from households and water treatment plants (Ternes et al., 2004).

3.7 Inorganic Contaminants

Inorganic contaminants include salts containing nitrogen (N), phosphorus (P), and potassium (K). Accumulation of these chemicals in water bodies is normally as a result of land use activities like fertilizer runoff or direct release from industry. Inorganic contaminants also include metals and metal particles which can be naturally occurring or as a result of runoff, industrial or domestic discharges, oil and gas production, mining, or farming. Most of these contaminant sources finally release into water bodies without proper treatment. On the other hand, industries such as dairy processing, mining, energy, and wastewater treatment can discharge wastewater that contain potential inorganic chemical contaminants (e.g., some metals such as mercury, copper, chrome, zinc, iron, arsenic, and lead bleach and curing agents).

3.8 Radioactive Contaminants

Radioactive contaminants can be naturally occurring or the result of oil and gas production, nuclear power plants, and mining activities. The radioactive particles in the water contain unbalanced number of protons and neutrons which are unstable in nature and can cause cancer. The radioactive waste can persist thousands of years in the environment, and the disposal of this waste is a great challenge around the world. The accidental release or improper disposal of contaminants threatens freshwater and marine water such as groundwater, surface water, and marine resources.

3.9 Biological Contaminants

The living organisms that can contaminate water are biological contaminants. Viruses, bacteria, and parasites that come from sewage treatment plants, septic tanks, agricultural livestock operations, and wildlife are the most common microbial contaminants in water and can cause diseases.

3.10 Sources of Contamination

A wide range of chemicals can contaminate our water and land impacting the environment and our health. Most contaminants enter the environment from industrial and commercial facilities; oil and chemical spills; nonpoint sources such as roads, parking lots, and storm drains; and wastewater treatment plants and sewage systems. Many hazardous waste sites and industrial facilities have been contaminated for decades and continue to affect the environment.

3.11 Sources of Contamination in Land

The sources of contamination can be classified based on spatiotemporal characteristics, nature of origin, and the quantity and type of contaminants released. A detailed classification by Fetter (1999) is as follows:

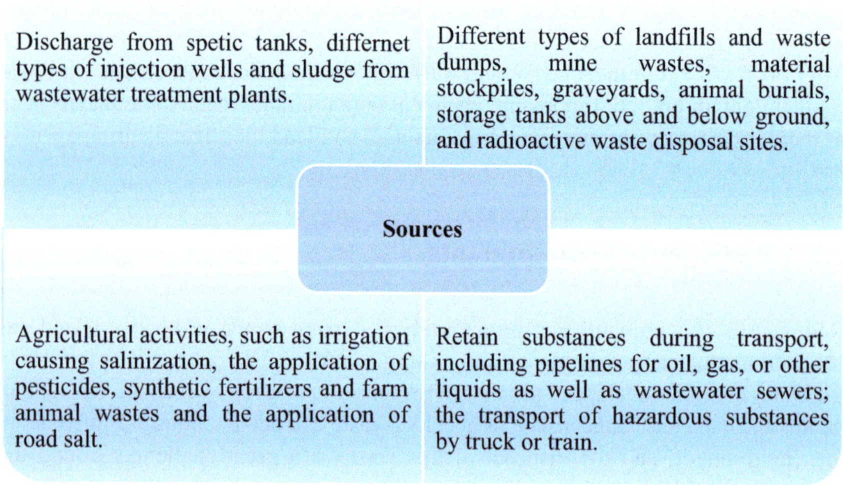

3.12 Sources of Contamination in Water

Generally, the contaminants can be released into the environment as gases and dissolved and particulate compounds in which it finally reaches to aquatic systems by various pathways such as atmosphere and soil. The contamination of water can come from three eminent sources:

1. Rainwater, the safest water to drink: It has become progressively contaminated in different areas of the world due to air pollution.
2. Surface water, in which runoff can pollute rivers, streams, and lakes: The major contamination comes from agriculture and industry.
3. Groundwater: The water from underground aquifers is normally safe to drink, and the water available in deep underground is more protected from many types of contamination.

But the geology and soil composition may affect water quality and be contaminated by naturally occurring elements like arsenic, selenium, fluoride, or boron.

The sources occur naturally, whose release is formed or repercussion by anthropogenic activities, which include surface water-groundwater interactions stimulated by pumping and saltwater intrusions due to overpumping of coastal aquifers (Fetter, 1999).

The sources of water contaminants can be point source and nonpoint source (Nathanson, 2022).

3.12.1 Point Source

If contamination is released from a single source which can be identifiable, it's called point source. The sources of discharge from industrial effluents and sewage treatment plant and contamination from leaking septic systems and chemical and oil spills are common examples of point sources. Although point source pollution initiates from a particular location, it can influence prolonged water bodies and ocean. The main point sources of contamination come from domestic waste discharge, industrial waste, and animal husbandry to freshwaters.

3.12.2 Nonpoint Source

Nonpoint source contamination is derived from diffuse sources, and then it is also called diffuse contamination. Nonpoint source contamination is associated with several sources such as agricultural runoff including soil erosion, waste, and precipitation. Contamination from diffuse sources is not easier to control than point sources.

3.13 Conclusion

Our land and water become strongly contaminated due to rapid industrialization and natural resource exploitation. Various types of contaminants such as natural or anthropogenic originate from different sources like point or nonpoint sources initially carried by different ways and finally reached to water bodies or contaminants dumped into the land by humans. Land and water contamination can therefore be a matter of concern in coming decades—maybe on account of rise in population, rapid urbanization, agricultural activities, and improper solid waste management. The occurrence of contaminants in land and water needs more attention, and the contamination of these resources should be prevented, controlled, and reduced to keep our sustainable environment.

Acknowledgments The author would like to express his sincere gratitude to the editors of this book for the support.

References

Akhtar, A. B. T., Naseem, S., Yasar, A., & Naseem, Z. (2021). Persistent organic pollutants (POPs): Sources, types, impacts, and their remediation. In R. Prasad (Ed.), *Environmental pollution and remediation. Environmental and microbial biotechnology*. Springer. https://doi.org/10.1007/978-981-15-5499-5_8

Cabeza, Y., Candela, L., Ronen, D., & Teijon, G. (2012). Monitoring the occurrence of emerging contaminants in treated wastewater and groundwater between 2008 and 2010. The Baix Llobregat (Barcelona, Spain). *Journal of Hazardous Materials, 239–240*, 32–39.

Doyle, P. (2021). Environmental geology. In: *Encyclopedia of geology*. Elsevier, pp. 660–668. https://doi.org/10.1016/B978-0-12-409548-9.12455-8

Fetter, C. W. (1999). *Contaminant hydrogeology*. Prentice Hall.

Knap, A. H., & Rusyn, I. (2016). Environmental exposures due to natural disasters. *Reviews on Environmental Health, 31*, 89–92.

Menon, N., George, G., Ranith, R., Sajin, V., Murali, S., Abdulaziz, A., Brewin, R. J. W., & Sathyendranath, S. (2021). Citizen science tools reveal changes in estuarine water quality following demolition of buildings. *Remote Sensing, 13*, 1683. https://doi.org/10.3390/rs13091683

Nathanson, J. A. (2022, March 3). *Water pollution*. Encyclopedia Britannica. https://www.britannica.com/science/water-pollution.

Nilsen, E., Smalling, K. L., Ahrens, L., et al. (2019). Critical review: Grand challenges in assessing the adverse effects of contaminants of emerging concern on aquatic food webs. *Environmental Toxicology and Chemistry, 38*(1), 46–60. https://doi.org/10.1002/etc.4290

Rhind, S. M. (2009). Anthropogenic pollutants: A threat to ecosystem sustainability? *Philosophical Transactions of the Royal Society, 364*, 3391–3401. https://doi.org/10.1098/rstb.2009.0122

Samaras, G. V., Stasinakis, S. A., Mamais, D., Thomaidis, S. N., & Lekkas, D. T. (2013). Fate of selected pharmaceuticals and synthetic endocrine disrupting compounds during wastewater treatment and sludge anaerobic digestion. *Journal of Hazardous Materials, 244–245*, 259–267.

Savita, Chopra. V., & Sharma, A. (2019). Environmental contaminants: Sources and effects. *Evaluation of Environmental Contaminants and Natural Products: A Human ealth Perspective, 1*, 1. https://doi.org/10.2174/9789811410963119010004

Ternes, A. T., Herrmann, N., Bonerz, M., Knacker, T., Siegrist, H., & Joss, A. (2004). A rapid method to measure the solid–water distribution coefficient Kd for pharmaceuticals and musk fragrances in sewage sludge. *Water Research, 38*, 4075–4084.

Vallero, D.A. (2010). Fundamentals of environmental chemistry. Environmental contaminants: Assessment and control. Elsevier. pp. 289–332. ISBN 9780080470351. Archived from the original on 2020-02-08. Retrieved 2019-04-12.

Yang, G. C. C., Yen, C. H., & C.L. (2014). Wang, monitoring and removal of residual phthalate esters and pharmaceuticals in the drinking water of Kaohsiung City Taiwan. *Journal of Hazardous Materials, 277*, 53–61.

Chapter 4
Contaminated Land/Water Microbial Diversity and Enzymatic Activity in Relation to Pollution

Muhammad Afzaal, Eman Idrees, Muhammad Farooq, Naeem Akhtar Abbasi, Hafiz Abdul Manan, Ahmad Farooq, Farida Begum, Shaukat Ali, Muhammad Huzaifa Mehboob, Gul Zareen Ghafoor, Muhammad Umar Hayat, Iqra Mazhar, Nusrat Bashir, Syeda Saira Iqbal, and Zafar Hashmi

4.1 Introduction

Pollutants are toxins that enter the environment beyond set limits and impact the people who live there. Pollutants can be chemical compounds or energy such as heat, sound, and light that come from sources. They may be naturally occurring or

M. Afzaal (✉) E. Idrees · M. Farooq · A. Farooq · M. H. Mehboob · G. Z. Ghafoor · M. U. Hayat · I. Mazhar · N. Bashir · S. S. Iqbal
Sustainable Development Study Centre, Government College University Lahore, Lahore, Punjab, Pakistan
e-mail: dr.afzaal@gcu.edu.pk; zareen.sdsc@gcu.edu.pk; umerhayyat@gcu.edu.pk

N. A. Abbasi
College of Earth and Environmental Science, University of Punjab, Lahore, Pakistan
e-mail: naeem.cees@pu.edu.pk

H. A. Manan · S. Ali
Department of Polymer Engineering, University of Punjab, Lahore, Pakistan
e-mail: hmannan.ipte@pu.edu.pk; dr.shaukat@kiu.edu.pk

F. Begum
Department of Environmental Science, Karakoram International University Gilgit, Gilgit, Pakistan
e-mail: farida.shams@kiu.edu.pk

Z. Hashmi
Department of Chemistry CIIT, Islamabad, Pakistan
e-mail: zaffar.hashmi@comsats.edu.pk

H. Alshemmari et al. (eds.), *Contaminated Land and Water*, Emerging Contaminants and Associated Treatment Technologies, https://doi.org/10.1007/978-3-031-65129-8_4

alien compounds to the ecosystem, such as metal pollution in mining areas, ash from volcanic eruptions, airborne ash deposition, particles, etc. Pollutants of many types, some of which may or may not be biodegradable, have been generated by rapid industrialization and natural resource depletion. Huge amount of industrial pollution discharges, heavy metals, organochlorine compounds, organophosphates, chlorinated hydrocarbons, insecticides, herbicides, and nitroaromatic compounds are significant environmental risks (Saravanan et al., 2021).

A contaminated site is any land or body of water that contains dangerous compounds in large amounts and is substantially likely to cause environmental damage, especially to human health. There may be contamination in the water or soil. Hotspots for pollution that are full of pollutants are classified as polluted sites. Although significant steps have been made in recent years to prevent industrial releases, releases still happen. Soil pollutants may diminish the biomass and some enzyme production of soil microbes, causing changes in soil communities.

Pollutants in soil and water are degraded, converted, and eliminated via a biological process known as bioremediation. It is a normal phenomenon in which bacteria, fungus, and plants degrade pollutants. These organisms' metabolic processes may use chemical pollutants as a source of energy, often producing innocuous or less harmful compounds (Singh et al., 2020). Microorganisms are commonly utilized in pollution management because they can degrade a wide range of organic compounds and absorb inorganic contaminants. In situ soil and groundwater bioremediation techniques, used for water treatment, are the most effective ways to use microbes (Fig. 4.1).

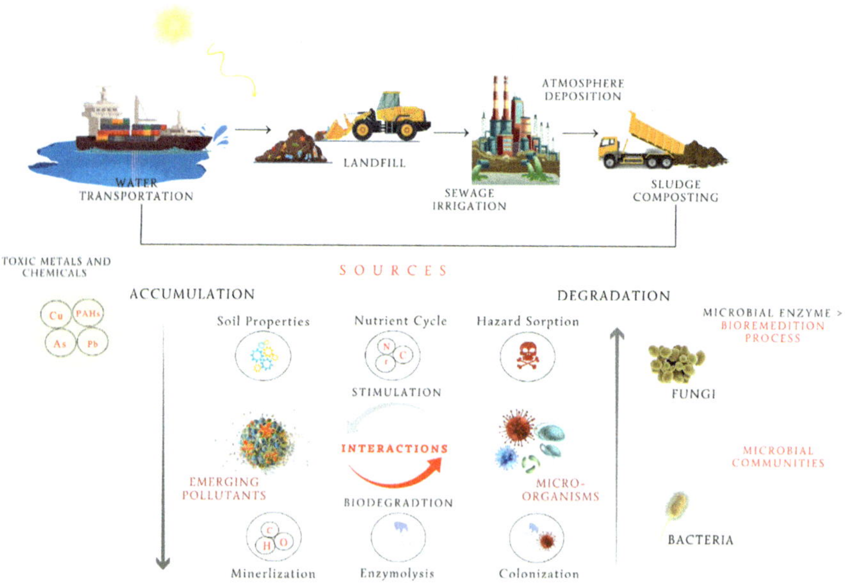

Fig. 4.1 Interaction between contaminated pollutants and microorganisms in the soil environment

4.2 Structure and Diversity of Soil Microbial Communities

Soil microorganisms drive up the majority of soil biodiversity and play an essential part in certain crucial soil ecosystem functions. Land use changes caused by humans influence the diversity and composition of the soil microbial community. Soil microorganisms help with inorganic matter conversion, nitrogen cycle, and xenobiotic chemical degradation. Soil microbes and enzymes are good markers of soil nutrients. They rapidly adjust accordingly, indicating metabolic alterations induced by pollution and toxins (Afzaal et al., 2018).

Soil microorganisms, microbial variety, and enzymatic activity are engaging markers for tracking soil disturbance or contamination because they play a fundamental and crucial part in soil ecosystem function. Since enzymes are susceptible to environmental stress and land-management practices, they could serve as prospective bioindicators' integrity for effective utilization. Enzymes are produced by organisms and play a vital role in microbial soil dynamics (Panneerselvam et al., 2022). Similar to this, many functional groupings of bacteria and fungi have also helped determine how anthropogenic causes have altered the structural makeup of microbial soil diversity.

Microorganisms are essential to soil ecology. Furthermore, there is a relationship between soil contamination and microbial diversity. The consistency of microbial diversity reflects the health of a microbial community and may be used to forecast the trend of ecological and soil nutrient uptake (Nikitin et al., 2022). Soil microbes play a vital function in the biogeochemical cycle system. It contributes to soil self-purification, hazardous chemical transition, and modification. Soil microorganisms, which are significantly more vulnerable to contaminants than soil creatures and plants, are indicators of alterations in the physical and chemical compositions of the soil and environmental protection.

4.3 Contaminated Water and Their Impact on Soil

4.3.1 Textile Industry

The textile industry released hazardous compounds into wastewater. The process of dyeing and finishing generates a lot of effluents. Among the trace elements found in improperly treated wastewater are cadmium, copper, arsenic, cobalt, lead, nickel, and mercury. Other contaminants include colors, surfactants, alcohol, cellulose, and high chemical and biological oxygen demands. The main threat to soil contamination is the release of sewage wastewater or the irrigation of contaminated wastewater.

4.3.2 Sewage-Contaminated Municipal Water

Irrigation of crops with sewage-contaminated water due to its low cost and high nutritional content. Irrigation with wastewater significantly affects the number of heavy metals in soil. The metal loading approach frequently causes soil deterioration and food chain contamination through the vegetables cultivated on such land (Khan et al., 2019). Farmers that irrigate their soils with sewage-tainted water have seen a decline in crop output. Heavy metals can seep into groundwater or remain and adsorb on soil particles. Similar to other environmental stresses, these metals also cause plants to produce more antioxidant enzymes.

4.3.3 Leather Manufacture

Large quantities of liquid and solid byproducts from leather manufacturing and tanneries pollute the Earth. The untreated tanning industry effluents are contaminated with chromium compounds, colors, chlorides, dissolved solids, nitrogen, and suspended particles. The tannery process is to blame for the number of incidents of chromium contamination in water bodies and soil (Taghipour et al., 2015). Chromium is heavily abundant in tannery effluents and effectively cleaned from the effluent might wind up in nearby water bodies and soils. Through soil irrigation from water contaminated with tanning effluents, tanning effluent-related soil contamination may also happen indirectly.

4.3.4 Industrial Waste

Industrial waste disposal facilities have contaminated the land of the rising industry. Every soil type has its personality. The mineral components in the discharged effluents, which are present in trace levels, encourage the development of certain bacterial, fungal, and algal colonies, which alter the soil's texture. Some of the pollutants deposited may also be absorbed by plants growing in contaminated soil. Organic wastewaters with high levels of released biodegradable organic materials that attract saprophytic soil flora can sometimes result in undesirable or fungal conditions due to their proliferating nature (Augustin et al., 2012). Industrial effluents discharged into the surroundings can lead to soil contamination. Highly hazardous chemical and industrial wastes when left untreated will severely damage the Earth. For example, the accumulation of radioactive material in deposits will cause contaminants to get discharged into the soil. This garbage has the potential to pollute groundwater. Chemical insecticides include a variety of hazardous compounds.

4.4 Relationship Between Enzymatic Activity and Pollution

Numerous studies have shown that soil pollution alters the enzymatic activity of bacteria and fungi. As a defense strategy, soil microorganisms can enhance the production of the phosphatase enzyme, which is used to phosphorylate substances and other pollutants, such as petroleum hydrocarbons, etc. This defensive strategy boosts baseline metabolism measured by production of CO2 while decreasing dehydrogenase activity because most of the carbon is consumed to produce these phosphorylated ligands rather than inorganic or organic substances that support growth and division of cells. One of the most significant enzyme groups in soil dehydrogenases is engaged in the recycling of organic matter in the soil, which has an impact on soil redox. Due to inhibition by pollutants such as trace elements, fertilizers, and pesticides, dehydrogenase activity has been identified as a promising bioindicator of soil pollution (Kuang et al., 2018). As a result of toxic metal contamination, community structure may change toward trace element-resistant microbial communities. Furthermore, nitrification enzymes are potential markers of pesticide impacts on soil microbes since the reaction to unfavorable effects of pollutants is due to the involvement of a few species in the nitrogen removal processes.

4.4.1 Heavy Metal Pollution

Heavy metal poisoning of the soil is a significant problem in the developed world. Heavy metal pollution not only has a detrimental influence on various aspects that affect plant quality and production, but it also changes the size, composition, and activity of microbial population (Hinojosa et al., 2004). Heavy metals are hence the source of soil pollution. Heavy metals are the root cause of soil pollution, particularly Ni, Cd, Cu, Zn, Pb, and Cr. The microbiological and biochemical features of soil were negatively impacted by heavy metals.

The characteristics of the soil, including its pH, clay content, and organic matter, have a considerable impact on the effects of elements on chemical and biological attributes. The production of enzymes by microbes is altered by heavy metals, which indirectly affects the enzymatic activity of soil. Through interfering with important bacterial activities and reducing the quantity and microbial activity, heavy metals have harmful effects on the soil biota. Heavy metal impacts might improve the tolerance of bacterial communities and fungi like arbuscular mycorrhizal (AM) fungi, which may be essential in the regeneration of damaged ecosystems (Chu et al., 2018).

In polluted soils, heavy metals resulted in reductions in both the productivity and variety of the microbial flora and an increase in soil actinomycetes. The enzymes catalysts in the soil have a variety of chemical preferences, and as a result, various metals have varying effects on enzyme activity. Cadmium is riskier to enzymes than lead due to its greater mobility and lower affinity for soil colloids. More so than cellulose activity, copper inhibits b-glucosidase. The activities of acid phosphatase, catalase, invertase, and urease are all significantly decreased by Pb. Phosphatase, arsenic (V), and sulfatase inhibit and urease is unaffected. Alkaline phosphatase, arylsulfatase, protease, and urease are negatively impacted by Cd contamination. However, invertase is not significantly affected. The toxic metals vary in each soil enzyme susceptibility. In general, the order of decreasing urease performance inhibition was Cr > Cd > Zn > Mn > Pb.

Indicators of soil nutrients, microbial diversity, and activity are provided by plant growth communities, which also play a vital role in nutrient cycling, soil health maintenance, toxic chemical detoxification, and pathogen management. It is crucial to look at how soil microbes in ecosystems subject to chronic heavy metal pollution function. Chromium is often present in soils such as Cr (III) and Cr (VI), each of which has unique chemical characteristics and toxicities. While Cr (VI) is a potent oxidizing agent and a dangerous species, Cr (III) is a micronutrient or a species that is 10 to 100 times less harmful than Cr(VI). Cr (VI) at high concentrations is recognized to have adverse effects on the metabolism of microbial cells and to alter the composition of bacterial populations in soil.

The metabolism of soil microorganisms was initiated to be negatively impacted by high amounts of metal (Cr, Zn, and Cd) contamination in all conditions. The contaminants have undesirable influences on bacteria, which changes the range, population, and records of the soil microbes. In general, an increase in metal concentration degrades the microbiological properties of the soil, including respiration rate and enzyme activity, which appear to be useful indicators of soil pollution.

4.4.2 Polycyclic Aromatic Hydrocarbons

Polycyclic aromatic hydrocarbons (PAHs), the most persistent organic pollutants, are very toxic to soil bacteria. Particularly in soil, high levels of PAHs may negatively affect the overall fungus and bacteria species, microbial metabolic functions, and enzyme activity. By placing pressure on subtle soil microorganisms, PAHs can modify the makeup of microbial communities. When microbes are exposed to PAH stress, their activity as measured by soil dehydrogenase and urease function is suppressed. Stress from PAHs reduced microbial richness and diversity (Picariello et al., 2023). *Significantly impacted by PAH stress, the bacteria Rhizobacter, Mycobacterium, Sphingobium, Bacillus, Pseudarthrobacter,* and *Massilia can be used as markers of PAH pollution in agricultural soils. Additionally, the bulk of microbial metabolic activities responds negatively to PAH stress. As a result, there was a noticeable shift in the metabolic content of the soil, which indicated a change in soil metabolism. In conclusion, the introduction of new contaminants into long-term contaminated soils hindered bacterial growth and metabolism, possibly having a significant impact on the overall soil quality (*Li et al., 2019*).*

4.4.3 Pesticides

The enormous community of microflora and many enzymes that catalyze different reactions are present in the soil. Pesticides change the enzymatic activities of the soil, including phosphatase (acid and alkaline), dehydrogenase, glucosidase, cellulase, urease, arylsulfatase, invertase, and amylase which are crucial for nutrient conversion and consequently have an impact on the fertility, quality, and productivity of the soil. The kind of pesticide used and its dosage determine how it affects enzymatic activity. Enzymes have a role in respiration, including dehydrogenase. The microbial population falls when applied a pesticide at a dangerous dose for the microorganisms, which also causes the dehydrogenase level to drop. Dehydrogenase activity, therefore, acts as a sign of any alterations in the microbial community. Quizalofop-p-ethyl has no impact on the enzyme; however, imazethapyr briefly lowers the dehydrogenase activity when used at indicated and half-recommended rates. The dehydrogenase activity is typically impacted when pesticides are used at their twofold recommended rate (Hassaan et al., 2020).

An enzyme called phosphatase removes the phosphoryl group. Acid phosphatase or alkaline phosphatase may be present depending on the pH of the soil. A few pesticides, such as quizalofop-p-ethyl and imazethapyr, have a transient impact on acid and alkaline phosphatase. Tembotrione, atrazine, pendimethalin, and topramezone initially raise the urease activity, which subsequently lowers the mixture of topramezone, and atrazine initially decreases the activity, which increases, then decreases, and eventually increases the activity (Kumari et al., 2018). A pesticide called Baythroid has an impact on the activities of amylase and invertase. High concentrations of pesticides such as chlorothalonil and propiconazole have a detrimental influence on the activities of cellulase and invertase.

4.5 Microbial Enzyme

Bacteria are widely spread throughout the biosphere and create enzymes due to their capacity for metabolism and ease of growth in different environmental circumstances. Many other enzymes from aerobic bacteria, including *Rhodococcus*, *Pseudomonas*, *Sphingomonas*, *Alcaligenes*, and *Mycobacterium*, have been known to break down pesticides and hydrocarbons. Bioremediation of trichloroethylene (TCE), chloroform, and polychlorinated biphenyls (PCBs) have all been using enzymes from anaerobic bacteria (Bhandari et al., 2021). Laccases, proteases, cytochrome P450, hydrolases, dehydrogenases, lipases, and dehalogenases are among the most common microorganism and genetically modified microorganism enzymes involved in the breakdown of the many types of pollutants used in bioremediation procedures.

4.5.1 Bioremediation Process

Pollutants are transformed into less hazardous or nonhazardous chemicals by the microorganisms that participate in bioremediation. It has been discovered that a wide range of organisms, including fungi, bacteria, algae, and plants, may be used in bioremediation (Fig. 4.2) (Chandran et al., 2020). Phytoremediation is a novel green approach for removing or degrading harmful substances from soils, groundwater, sediments, and surface water. Bioremediation is based on organisms that attack pollutants enzymatically and convert them to harmless chemicals (Borah, 2018).

Bioremediation is a significant technique. Only a few fungus and bacteria are efficient polluters. Because bioremediation is only effective when environmental conditions permit microbial proliferation and activity, its application typically requires modifying parameters to allow for faster microbial growth and destruction (Pande et al., 2020). Bacteria and fungus both rely on the activation of many intracellular or extracellular enzymes.

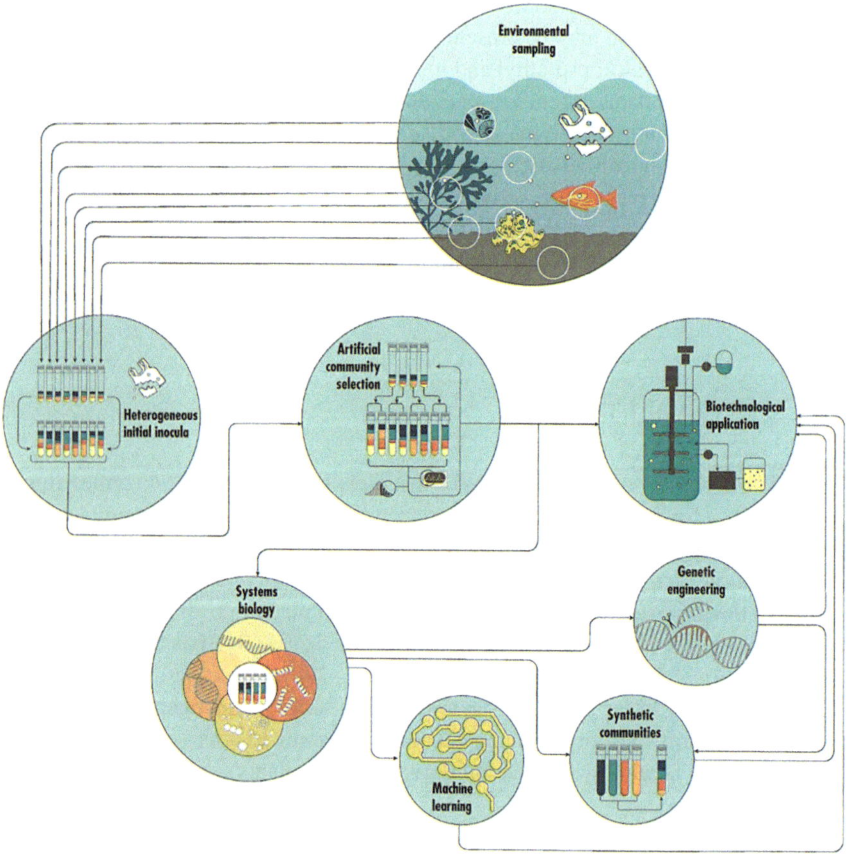

Fig. 4.2 A plan for using artificial community selection for pollution bioremediation by microorganisms

4.6 Bioremediation of Pollutants Using Microbial Enzymes

4.6.1 *Enzymes for Organic Substrates*

Large amounts of environmental contaminants, such as pesticides, medicines, plastics, and dyes, pollute the land and water every year. Aromatic compounds, organ cyanides, steroids, PAHs, polymers, and chlorinated hydrocarbons are among the organic contaminants that must be cleaned up globally. Their rigid composition is the fundamental reason of their toxicity.

Microbial Hydrolase
Various substances, including organophosphorus compounds, herbicides, nitrile compounds, insecticides, and polymers, can be remediated using hydrolase enzymes. To control waste generated during food processing, degrade plastics and pesticides, remove biofilm deposits, and clean up oil-contaminated soils, among other things, microbial hydrolytic enzymes are employed. Hydrolytic enzymes have possible applications in the chemical, feed, and medicinal sectors (Eibes et al., 2015). Amidase, esterase, and protease activity may cause the breakdown of amide, ester, and peptide bonds, resulting to a chemical that has minimal or no toxicity being formed. To convert pollutants like carbofuran, diazinon, carbaryl or parathion, and coumaphos through the process of hydrolysis, microbial hydrolase from *Pseudomonas*, *Achromobacter*, *Flavobacterium*, *Bacillus*, *Nocardia*, and *Cereus* has been utilized effectively (Fig. 4.3).

Microbial Laccase
The enzyme benzenediol oxygen oxidoreductase, often known as laccase, is useful for oxidizing phenolic compounds, dyes, insecticides, and PAHs. Among the following processes occurs (Fig. 4.4) when a laccase substrate acts as an oxidase (1) breakage of aromatic rings, (2) polymerization, or (3) breakdown of covalent connections among monomers (Chauhan et al., 2017). Various fungal species, including *Polyporus* sp. and *Panus conchatus*, are where laccase was initially identified. Later, *Azospirillum lipoferum*, the first bacterium species, was discovered to have laccase.

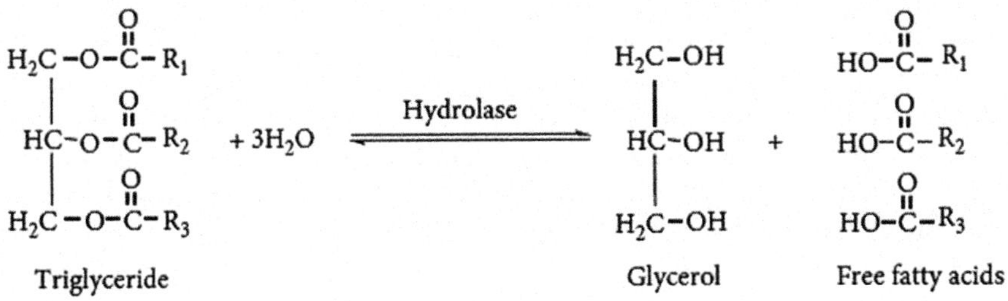

Fig. 4.3 Enzymatic reactions catalyzed by Hydrolase

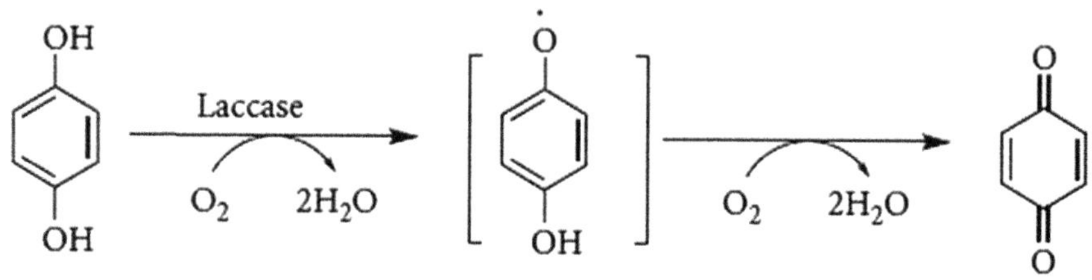

Fig. 4.4 Enzymatic reactions catalyzed by laccase

$$RH + O_2 + NAPH + H^+ \xrightarrow[Fe^{3+}]{\text{Cytochrome P450}} ROH + H_2O + NAP^+$$

Fig. 4.5 Enzymatic reactions catalyzed by Cytochrome P450

Laccase is generated in a range of Gram-positive bacteria, with *Lysinibacillus*, *Geobacillus*, *Bacillus*, *Aquisalibacillus*, *Staphylococcus*, and *Streptomyces*. Numerous bacteria can release laccase outside their cells, although some cannot. Extreme temperatures and pH levels don't affect bacterial laccase as much.

Microbial Monooxygenase
A class of heme-containing enzymes known as cytochrome P450 monooxygenases (CYP) catalysis many processes including N-dealkylation, N-hydroxylation, oxidative dehalogenation, and O-dealkylation. NADPH-CYP reductase, an enzyme that obtains its electrons from atmospheric oxygen, provides CYP with the necessary electrons for processes (Fig. 4.5). Therefore, it is required to have a reducing agent like NAD(P)H or FAD (Li et al., 2020). CYPs are versatile enzymes found in a variety of bacterial, fungal, plant, and animal species. Among the genera whose cytochrome P450 is utilized for bioremediation contain *Streptomyces*, basidiomycete, *Bacillus*, *Dehalococcoides*, *Saccharomyces*, *Rhodococcus*, *Salmonella*, and *Escherichia*.

4.6.2 *Enzymes for Inorganic Substrates*

Most microbes create metal-binding peptides like metallothioneins and phytochelatins when poisonous heavy metals are present, which reduces their toxicity by sequestering them. For instance, the enzyme phytochelatin synthase creates phytochelatin, which works with GSH to collect heavy metals. Numerous microbes evolved particular heavy metal resistance mechanisms. Enzymes are undoubtedly the significant component of these routes; the following is an overview of a few of these metal-specific enzymes (Pratush et al., 2018):

Arsenic

Arsenic is a heavy element found in nature in both biological and manufactured forms. The inorganic forms (As3+ and As5+) are toxic and could inhibit enzyme activity in soil. Arsenic is treated by many different microorganisms, like *Pseudomonas* sp., *Acinetobacter* sp., and *Sporosarcina ginsengisoli.* Engineered species for As bioremediation include *Bacillus idriensis*, *E. coli*, and *Sphingomonas desiccabilis* (Sher et al., 2019). Some fungi, such as *Rhizobium* sp., *Trichoderma* sp., *Penicillium canescens*, *Aspergillus flavus*, and others, are also arsenic bioremediators. The ability for bioremediation occurs in plant *Pteris vittata* of the Pteridaceae family and in the algal genera, *Rhizoclonium*, *Hydrodictyon*, and *Oedogonium* (Sher et al., 2019).

Lead

Before industrialization, lead was present in trace amounts in nature. However, other Pb compounds now arise in and pollute the soil through gasoline combustion. However, through bioremediation processes, certain bacteria may break down organ lead (Naik et al., 2013). *Biotransforming organisms include Rhizopus nigricans, Saccharomyces cerevisiae, Penicillium canescens, Agaricus bisporus, Penicillium chrysogenum, Aspergillus niger, and Aspergillus terreus. Furthermore, Phaeolus schweinitzii and Arthrobacter are known to decompose trimethyl lead cations.*

Chromium

Because of its great oxidative capacity, Cr (VI) is considered the most toxic heavy metal. The widespread utilization of chromium and its derivatives and mining contribute to the pollution of soil and water. Hexavalent chromium is bioremediation by being reduced to trivalent species. Some species sensitive to Cr (VI) and may lower it are *Bacillus*, *Pseudomonas*, *Escherichia*, *Enterobacter*, *Shewanella*, and *Thermus*. Hexavalent chromium degradation can take place via anaerobic or aerobic mechanisms. The enzymes with hexavalent chromium-reducing ability are Ni-Fe dehydrogenase, chromate reductase, and cytochrome c3 (Thatoi et al., 2014).

4.7 Conclusion

Water pollution caused by human-related activities has resulted in the production of pollutants that are harmful to the environment. It decreases soil microbial biomass and various enzyme reactions as soon as it enters the soil environment. Soil bacteria have a substantial impact on soil health and quality. They are critical for sustaining soil fertility for the organic matter cycle, nutrient release, and soil structure stabilization. Heavy metals, antibiotics, pesticides, toxic hydrocarbons, and other natural and man-made pollutants can both disrupt soil balance. By altering the microbial composition of the land, the antibacterial effect of these compounds can differently prevent the growth of soil microbes, which can modify the ecological function of the soil. Due to their quick reactivity and susceptibility to contaminants, microbial parameters are valuable indicators of soil fertility and qualitative condition.

References

Afzaal, M., Mukhtar, S., Malik, A., Murtaza, R., & Nazar, M. (2018). Paddy soil microbial diversity and enzymatic activity in relation to pollution. In *Environmental pollution of Paddy soils* (pp. 139–149). Springer.

Augustin, P. V., & Viero, A. P. (2012). Environmental impact and geochemical behavior of soil contaminants from an industrial waste landfill in Southern Brazil. *Environmental Earth Sciences, 67*(5), 1521–1530.

Bhandari, S., Poudel, D. K., Marahatha, R., Dawadi, S., Khadayat, K., Phuyal, S., Shrestha, S., Gaire, S., Basnet, K., & Khadka, U. (2021). Microbial enzymes used in bioremediation. *Journal of Chemistry, 4*, 1–17.

Borah, D. (2018). Microbial bioremediation of petroleum hydrocarbon: An overview. *Microbial Action on Hydrocarbons*, 321–341.

Chandran, H., Meena, M., & Sharma, K. (2020). Microbial biodiversity and bioremediation assessment through omics approaches. *Frontiers in Environmental Chemistry, 1*, 570326.

Chauhan, P. S., Goradia, B., & Saxena, A. (2017). Bacterial laccase: Recent update on production, properties and industrial applications. *Biotech, 7*(5), 1–20.

Chu, L., & Dian. (2018). *Effects of heavy metals on soil microbial community*. IOP Conference Series (Vol. 113). Earth and Environmental Science.

Eibes, G., Arca-Ramos, A., Feijoo, G., Lema, J., & Moreira, M. (2015). Enzymatic technologies for remediation of hydrophobic organic pollutants in soil. *Applied Microbiology and Biotechnology, 99*(21), 8815–8829.

Hassaan, A. M., & El Nemr, A. (2020). Pesticides pollution: Classifications, human health impact, extraction and treatment techniques. *The Egyptian Journal of Aquatic Research, 46*(3), 207–220.

Hinojosa, M. B., Carreira, J. A., García-Ruíz, R., & Dick, R. P. (2004). Soil moisture pre-treatment effects on enzyme activities as indicators of heavy metal-contaminated and reclaimed soils. *Soil Biology and Biochemistry, 36*(10), 1559–1568.

Khan, Z. I., Ahmad, K., Rehman, S., Ashfaq, A., Mehmood, N., Ugulu, I., & Dogan, Y. (2019). Effect of sewage water on accumulation of metals in soil and wheat in Punjab, Pakistan. *Pakistan Journal of Analytical & Environmental Chemistry, 20*(1), 60–66.

Kuang, S., Su, Y., Wang, H., Yu, W., Lang, Q., & Matangi, R. (2018). Soil microbial community structure and diversity around the aging oil sludge in yellow river delta as determined by high-throughput sequencing. *Archaea, 2018*.

Kumari, J. A., Rao, P., Madhavi, M., & Padmaja, G. (2018). Effect of herbicides on the activity of soil enzymes urease in maize crop. *Indian Journal Of Agricultural Research, 52*(3).

Li, X., Qu, C., Bian, Y., Gu, C., Jiang, X., & Song, Y. (2019). New insights into the responses of soil microorganisms to polycyclic aromatic hydrocarbon stress by combining enzyme activity and sequencing analysis with metabolomics. *Environmental Pollution, 255*, 113312.

Li, Z., Jiang, Y., Guengerich, F. P., Ma, L., Li, S., & Zhang, W. (2020). Engineering cytochrome P450 enzyme systems for biomedical and biotechnological applications. *Journal of Biological Chemistry, 295*(3), 833–849.

Naik, M. M., & Dubey, K. S. (2013). Lead resistant bacteria: Lead resistance mechanisms, their applications in lead bioremediation and biomonitoring. *Ecotoxicology and Environmental Safety, 98*, 1–7.

Nikitin, D., Semenov, M., Chernov, T., Ksenofontova, N., Zhelezova, A., Ivanova, E., Khitrov, N., & Stepanov, A. (2022). Microbiological indicators of soil ecological functions: A review. *Eurasian Soil Science, 55*(2), 221–234.

Pande, V., Pandey, S. C., Sati, D., Pande, V., & Samant, M. (2020). Bioremediation: An emerging effective approach towards environment restoration. *Environmental Sustainability, 3*(1), 91–103.

Panneerselvam, P., Mitra, D., Mohanty, B., Mondal, R., & Shadangi, S. (2022). *Soil pollution and their impact on soil microorganisms*. AZRA India.

Picariello, E., Baldantoni, D., & De Nicola, F. (2023). How soil microbial communities from industrial and natural ecosystems respond to contamination by polycyclic aromatic hydrocarbons. *PRO, 11*(1), 130.

Pratush, A., Kumar, A., & Hu, Z. (2018). Adverse effect of heavy metals (As, Pb, Hg, and Cr) on health and their bioremediation strategies: A review. *International Microbiology, 21*(3), 97–106.

Saravanan, A., Kumar, P. S., Vo, D.-V. N., Jeevanantham, S., Karishma, S., & Yaashikaa, P. (2021). A review on catalytic-enzyme degradation of toxic environmental pollutants: Microbial enzymes. *Journal of Hazardous Materials, 419*, 126451.

Sher, S., & Rehman, & Abdul. (2019). Use of heavy metals resistant bacteria—A strategy for arsenic bioremediation. *Applied Microbiology and Biotechnology, 103*(15), 6007–6021.

Singh, S., Kumar, V., Datta, S., Dhanjal, D. S., Sharma, K., Samuel, J., & Singh, J. (2020). Current advancement and future prospect of biosorbents for bioremediation. *Science of the Total Environment, 709*, 135895.

Taghipour, M., & Jalali, & Mohsen. (2015). Effect of clay minerals and nanoparticles on chromium fractionation in soil contaminated with leather factory waste. *Journal of Hazardous Materials, 297*, 127–133.

Thatoi, H., Das, S., Mishra, J., Rath, B. P., & Das, N. (2014). Bacterial chromate reductase, a potential enzyme for bioremediation of hexavalent chromium: A review. *Journal of Environmental Management, 146*, 383–399.

Chapter 5
Impact of Biochar on the Pollutants in Contaminated Land and Fertility

Muhammad Afzaal, Muhammad Umer Farooq Awan, Shahzad Sharif, Muhammad Umar Hayat, Muhammad Luqman, and Zafar Hashmi

5.1 Introduction

Soil pollution is now a worldwide issue because of its adverse impact on the environment and health hazards. Soil contamination caused by natural or artificial pollutants is a widespread problem. The prolonged application of insecticides, effluent discharge, fertilizers, wastewater irrigation, and other practices contaminated agricultural soils (Murtaza et al., 2021). Organic pollutants, like polychlorinated biphenyls (PCBs), phthalate esters (PAEs), organochlorine pesticides (OCPs), and polycyclic aromatic hydrocarbons (PAHs), are distinguished in the environment by their persistence, toxic effect, and bioavailability (Sun et al., 2018). Soil

M. Afzaal (✉) · M. U. Hayat
Sustainable Development Study Centre, Government College University Lahore, Lahore, Punjab, Pakistan
e-mail: dr.afzaal@gcu.edu.pk; umerhayyat@gcu.edu.pk

M. U. Farooq Awan
Department of Botany, Government College University, Lahore, Pakistan
e-mail: Dr.umerfarooqawan@gcu.edu.pk

S. Sharif
Department of Chemistry, Government College University, Lahore, Pakistan
e-mail: mssharif@gcu.edu.pk

M. Luqman
Department of Environmental Sciences, University of Veterinary and Animal Sciences, Lahore, Pakistan
e-mail: muhammad.luqman@uvas.edu.pk

Z. Hashmi
Department of Chemistry CIIT, Islamabad, Pakistan
e-mail: zaffar.hashmi@comsats.edu.pk

H. Alshemmari et al. (eds.), *Contaminated Land and Water*, Emerging Contaminants and Associated Treatment Technologies, https://doi.org/10.1007/978-3-031-65129-8_5

contamination has damaging impacts on the environment, such as reducing fertility, destroying vegetation, and increasing soil erosion.

In recent years, increased focus has been placed on the treatment of soil pollution and potential environmental harm. There is an immediate need to determine ecologically friendly, economically viable, and suitable cleanup procedures. Biological chemical or physical treatments are currently the most vital technique. For the remediation of polluted sites, methods such as photocatalytic degradation, electrochemical degradation, bioremediation, and adsorption have been examined. Some approaches are challenging and perhaps problematic to use in soil cleanup with substantial contamination because of their lengthy periods and low efficacy (Issaka et al., 2022). Traditionally, pollutants are removed from the biosphere using chemical solvent extraction. Nevertheless, a number of issues, including high prices, soil degradation, nutrient loss, or massive environmental hazards, limit the use of efficient technologies in vast agricultural areas.

Adsorption is an inexpensive and energy-efficient approach for cleaning up contaminants (Zhang et al., 2018). A by-product of biomass pyrolysis is biochar. In recent years, biochar has received progressively more interest because of its potential use in restoring damaged soil and enhancing soil fertility. Through adsorption or bioremediation of the biochar, organic materials are concentrated by the hydroxyl groups and then broken down by the microorganisms. Biochar can clean up biologically polluted soil. It could also clean up inorganic pollutants through metal ions, electrostatic forces, and precipitation. Biochar's remediation action is thoroughly associated with its physical and chemical characteristics, influenced by the pyrolytic parameters such as temperature or heating rate and sources of raw materials like biosolids and lignocellulosic biomass impacted by soil conditions, including ionic strength, pH, and humidity (Wang et al., 2020).

5.2 Source of Pollutants in Land and Fertility

Numerous substances can contaminate soils, including direct inputs (point sources of pollution) like fertilizer and pesticide application as well as indirect inputs (diffuse pollution) like air deposition and flooding. Additionally, contaminated soils are a secondary source of emissions of pollutants into the air, surface waters, groundwater, and eventually seas. Pesticides, mineral fertilizers, organic fertilizers (manure and sewage sludge), wastewater for irrigation, plastic materials (such as films for mulching and greenhouses, drip irrigation tubes, and empty packaging), and rural wastes are the main sources of soil pollution in agricultural areas that affect the fertility of the soil in a specific area (Mirsal, 2008).

Table 5.1 shows that soil is among the essential resources humans rely on for survival, and it serves as the primary storehouse for biochemical processes. The technical group has made remediation of polluted soils a priority of safeguarding population well-being and promoting long-term sustainability. Soil pollution must be remedied, maintain the health of individuals, and ensure sustainability. Pesticides,

Table 5.1 Pollutant sources and their effects on soil fertility

SR no.	Pollutant	Sources	Impact on soil	Reference
1.	Heavy metals	Leaching from landfills Industrial effluents Human anthropogenic activities	Affects soil fertility Affects soil microbiota Reduces plant growth Reduces yield by accumulation	Nyiramigisha (2021)
2.	Antibiotic residues	Animal manure Fertilization Human waste	Influence soil microorganisms, modifying their enzyme activity and changing the total microbial biomass	Cycoń et al. (2019)
3.	Plastics	Plastic containers Food packing bags Plastic drink bottles	Fertility of the soil and chlorinated plastics have an impact on soil	Ansari et al. (2022)
4.	Pesticides	Pesticide industries Rainwater runoff Improper pesticide disposal	Reduces soil biodiversity, influences nontarget species and the possibility of food poisoning	Gill and Garg (2014)
5	Polycyclic aromatic hydrocarbons (PAHs)	Petrochemical refineries Gasification of coal Automobile exhaust	Decreases soil quality, reduces soil biodiversity, and results in genetic mutation	Yadav et al., 2021

polycyclic aromatic hydrocarbon (PAH), heavy metals, antibiotic resistance genes (ARGs), and antibiotic residues are soil pollutants that almost affect the environment and health.

5.3 Remediation of Soil Contamination

5.3.1 Biochar

Biochar is an activated carbon depending on the temperature used for decomposition and the kind of feedstock utilized. The biochar generated has different physical and chemical properties. It might be applied to lessen the metals' availability and capacity to leach from polluted soils. Additionally, it can enhance the quality of polluted soil and significantly lessen heavy metals that particular soil can absorb (Hossain et al., 2020). Additionally, it is a substance that contains biomass, which minimizes waste. It can combat climate change, generate biofuel, and improve soil fertility. Thus, biochar use can be a spectacular remedy in addition to cleanup polluted soil but also dealing with the challenges of climate change, exhaustion of fossil fuels, and food scarcity.

Biochar Production: Biochar is formed when raw material is thermally conserved. Pyrolysis, hydrothermal carbonization, and gasification methods are generally used in the process (Xie et al., 2015). Lately, application of more synthetic strategies has been increased to work on the properties of biochar. Different types of agricultural waste are selected as raw materials, dried by ordinary air, quickly crushed, filtered, and converted carbon in the burner. The biochar is dry after being cleaned with distilled water to remove surface debris or impurities (Chen et al., 2021). As of its spongy design and various practical gatherings, biochar has gotten consideration for its connection to the adsorption and immobilization of organic contaminants in soil. Nonetheless, the effectiveness of biochar is enormously reliant upon its assembling influences, which take into account various kinds of feedstock, manufacturing processes, and handling circumstances, all of which affect the cycles.

As shown in Table 5.2. It is indicated by the excess categories biochar can introduce different physiochemical assets of biochar's substance of the unpredictable matter, breaking down natural matter, debris, and carbon. Furthermore, the pyrolysis form might impact biochar's physicochemical and synthetic properties. Some qualities are usually present, for example, a broad surface area, a stable structure, a strong capacity for cation exchange, and a rich C/N ratio (Wang & Shizong., 2019). These attributes, in the particular unambiguous surface region, porosity, and cation trade limit, are answerable for the high adsorption limit that permits it to eliminate natural contaminations and weighty metals. Likewise, the arrangement pH affects the adsorption limit of biochar. Besides, biochar will work on the dirt's natural exercises, supplement maintenance, water maintenance limit, increment of pH rate, and measuring natural soil matter (Guo et al., 2020).

Table 5.2 An overview of a few current research using biochar and its effects on pollutants

SR no.	Amendment type	Contaminants	End point	Efficiency	Reference
1.	Hardwood-derived biochar	PAHs	Earthworm bioaccumulation	45%	Kołtowski et al. (2016)
2.	Wheat ash	Diuron	Sorption	75%	Quirantes et al. (2017)
3.	Charcoal-derived biochar	Terbuthylazine	Sorption	80%	Gámiz et al. (2019)
4.	Orchard prune residue-derived biochar	Cd, Cr, Cu, Ni, Pb, and Zn	Sorption	95%	Ali et al. (2020)
5.	*Eucalyptus* spp.-derived biochar	Chlorpyrifos and carbofuran	Microbial degradation	65%	Deolikar & Patil (2022)

5.4 Heavy Metals

Heavy metal soil pollution is now a significant environmental and public health hazard. Its excessive concentration in the soil has dangerous impacts on productivity, the security of the food supply, and people's health. The quantity and type of metal exposure, the soil's characteristics, and the speed and extent of root uptake all affect the absorption of land toxic substances entering the food web (Awad et al., 2021).

The primary goal of scientists or environmentalists would be to prevent metals from entering the food chain to improve the health of individuals. A wide variety of toxic substances have accumulated significantly in the soil by irresponsible waste disposal methods (He et al., 2019). Anthropogenic activities with sewage drainage, vehicle emissions, mining or melting, and sludge treatments might be causes of soilborne heavy metals. Heavy metals may affect soil richness, microbiological activity, diversification, agricultural production, and health hazards by food consumption (Bolan et al., 2018).

Biochar offers a more effective, quicker, or eco-friendly replacement for toxic substances in soil samples. Biochar can treat soil that has been affected by toxic substances. The main processes that lower heavy metal toxicity in soils include ion exchange, precipitation, and adsorption (Cheng et al., 2020). Numerous studies show that biochar is beneficial in treating soils with heavy metals. Accordingly, research done by Zeng et al. indicated maize straw biochar decreased the bioavailability of cadmium in soil by changing it into a condition of reduced accessibility (Zeng et al., 2018). Furthermore, tests conducted by Chagas et al. utilizing sewage sludge biochar proved successful at reducing nonessential toxic substances inert for plants (Chagas et al., 2021). Li et al. used soybean straw biochar (3%) over cadmium- and arsenic-polluted agricultural soil, which decreased arsenic bioavailability in the wheat plant by about 88% and had a similar impact on cadmium (Li et al., 2018).

5.4.1 Antibiotics

Antibiotic overuse has resulted in irregular waste release in the ecosystem, causing several environmental issues. Due to the possibility that microbial contamination in the environment could harm nature, environmental poisoning of antibiotics would be a topic of increasing concern (Krasucka et al., 2021). Antibiotic metabolites have been identified in food, dung, surface waterways, and even soil or vegetation. As a result, antibiotics degrade slowly in nature and can remain for prolonged periods.

The biological method cannot adequately remove antibiotics. Adsorption is among the essential techniques regulating the destiny, transport, and responsiveness of antibiotics, regardless of the various dispersion paths for antibiotics in dung. Adsorption plays a crucial role in removing drugs from the ecosystem in particular

(Guo et al., 2020). Highly adsorbent biochar is effective in reducing the consequences of antibiotic remains in cow waste, which spread after cow manure is put into the soil. Biochar also can absorb environmental organic pollutants. Numerous antibiotics, particularly ciprofloxacin, fluoroquinolone, tetracycline hydrochloride, oxytetracycline, and doxycycline hydrochloride, are strongly absorbed by biochar. According to Pan et al., biochar rice straw eliminates tetracyclines and associated metabolites by changing soil characteristics and boosting antibiotic bioavailability (Pan & Min., 2020).

5.4.2 Pesticides

Pesticides are any product or combination of compounds intended to stop, eradicate, resist, or mitigate any pest (insects, weeds, animals, or microbes) as a pesticide. Pesticides are frequently employed in farming technology to stop or slow the development of crop pests. Ecosystem and health impacts are in danger from chemical discharge or deposits in the groundwater and soil (Varjani et al., 2019). Due to their resistive, chronic, and toxin characteristics and their capacity to travel great distances and damage far-off, populated regions of the world, pesticides constitute among the persistent organic contaminants. By providing access to carbon and other resources, biochar stimulates bacterial growth, which speeds up the anaerobic decomposition of pesticide present in the soil (Morillo & Villaverde, 2017).

Numerous kinds of research have been performed to examine the impact of using biochar for pesticide cleanup. Pesticides such as atrazine, pyrimethanil, terbuthylazine, and bentazone were successfully absorbed by biochar. It was demonstrated that using charcoal causes insecticides to permanently bind to soils due to splitting into compacted formations, surface-specific binding, or trapping in biochar small pores (Wang et al., 2010). Biochar microporosity and mesoporosity were the main factors that influenced triazine herbicide adsorbed by biochar. As for purpose of atrazine adsorption, Liu et al. likened biochars resulting from manure (pig, poultry, and cattle) and plants (corn, rice stalks, and soybean) and came to the conclusion that, in overall, plant-related biochars were more efficient at doing so. The percentage removal also enhanced by n rise in high heat remedies as well as biochar pH (Liu et al., 2015).

In amended soil, examining the ability of biochar to absorb sulfamethazine by biochar made from an exotic cucumber plant *Sicyos angulatus* L. in soil containing 5 mg kg1 of sulfamethazine pesticide, it was observed sulfamethazine decreased 85% behind the use of biochar. However, with 50 mg kg1 of pesticide in the soil, the sulfamethazine was reduced by 63%. Vithanage et al. utilized cucumber biochar (Sicyos angulatus L.) to remove sulfamethazine from sandy and clay soils below various pH values and sulfamethazine conc. Biochar pyrolyzed at an elevated temp (700 °C) displayed a high level of sulfamethazine absorption. At 3.0 pH, sulfamethazine absorption extended its highest and is influenced by electrostatic interactions. The primary sorption mechanism cation exchanged the increased pH 5.0

to7.0. In sandy and clay soil, biochar was capable of keeping increases in sulfamethazine absorption of up to 80% to 82% (Vithanage et al., 2014). Large-scale field experimentation is necessary before remediation initiatives are in use because of biochar treatment investigations for polluted soil cleanup in labs, nurseries, or particular plot trials. In overview, the pyrolysis process impacts the characteristics of biochar and pesticide adsorption ability.

5.4.3 Antibiotic Resistance Genes

Antibiotic resistance genes (ARGs) become the mechanisms by which microorganisms develop extreme levels of antibiotic resistance. Antibiotic resistance genes (ARGs) have been receiving increased attention as one of the main dangers to agricultural production and worldwide health disease over the past decade (Hernando-Amado et al., 2019). One form of developing pollutant is the growth of microorganisms immune to antibiotics, which is becoming an increasing worldwide issue. Animal manure soil amendments are possible sources of ARGs in soil and plants. The widespread application of animal waste to the land encourages the transmission of antibiotic-resistant bacteria, mostly from soil to people through the food supply chain.

Since biochar modifications may influence how ARGs spread and finish up in the natural environment, reducing their quantity, some surveys also investigated the application of biochar to reduce soil ARG contamination. Numerous investigations revealed that biochar effectively reduced ARG contamination in soils to a certain level. According to research conducted by Chen et al., the biochar rice straw method was evaluated to diminish the richness of 131 ARGs in unplanted soil, but a negligible impact was observed in soil which had been planted with *Brassica chinensis* L. In this experiment, 0.6% biochar was added to the soils as an alteration or not. The studies demonstrate that the addition of disintegrated biochar significantly reduced the occurrence of recombinant ARGs between bacteria. This research also shows that biochar interaction significantly reduced the occurrence of ARG multiplication (Chen et al., 2018).

According to Cui et al., ARG Tet and sul genes were shown to be more prevalent in surrounding soil after being added to biochar wheat straw. These investigations demonstrated the unpredictability of biochar's efficacy in removing soil ARG contamination. Such an assumption is linked to the characteristics of the biochar since these characteristics, specifically raw material and pyrolytic circumstances, may alter the recombinant transmission of ARGs among bacteria (Cui et al., 2018). Additionally, the quantity of toxic substances or drugs in the soil affects how ARGs evolve in soil, making the production of biochars with superior heavy metal and antibiotic adsorption ability to prevent ARG soil contamination intriguing. The relationship between biochar and ARGs is substantially understudied than that among heavy metals and many organic contaminants. Thus, additional research is required to evaluate the impact of various biochar on ARGs and propose alternative methods to increase biochar's effectiveness in removing ARG-related land pollution (Lian et al., 2020).

5.5 Conclusion

Biochar is commonly applied in soil remediation and has shown substantial potential for reducing the adverse effects of PAHs, antibiotic resistance genes, heavy metals, antibiotics, and pesticides. The effectiveness of biochar in decreasing the amount and transportation of toxic compounds in soil depends on the characteristics and composition of the pollutants. Along with the conditions of the adsorption process, it depends upon the physicochemical characteristics of biochar. The type of feedstock utilized and the heat of the pyrolysis are the main determinants of biochar qualities. Therefore, it is crucial to choose the feedstocks and manufacturing settings to produce biochar adsorption that is effective and selective. In this regard, future research should concentrate on the outcome of inexpensive biochar amendments with great real applications and significance on ecological fate on agricultural fields. Alternate uses for modified biochar include carbon sequestration, improving soil fertility, cleaning up pollutants, and reusing waste materials and by-products.

References

Ali, A., Shaheen, S. M., Guo, D., Li, Y., Xiao, R., Wahid, F., & Zhang, Z. (2020). Apricot shell--and apple tree-derived biochar affect the fractionation and bioavailability of Zn and Cd as well as the microbial activity in smelter contaminated soil. *Environmental Pollution, 264*, 114773.

Ansari, A. A., Naeem, M., Gill, S. S., & Siddiqui, Z. H. (2022). Plastics in the soil environment: An overview. *Agrochemicals in Soil and Environment: Impacts and Remediation*, 347–363.

Awad, M., Liu, Z., Skalicky, M., Dessoky, E. S., Brestic, M., Mbarki, S., Rastogi, A., & El Sabagh, A. (2021). Fractionation of heavy metals in multi-contaminated soil treated with biochar using the sequential extraction procedure. *Biomolecules, 11*(3), 448.

Bolan, N., Kunhikrishnan, A., Thangarajan, R., Kumpiene, J., Park, J., Makino, T., Kirkham, M. B., & Scheckel, K. (2018). Remediation of heavy metal (loid) s contaminated soils–to mobilize or to immobilize? *Journal of Hazardous Materials, 266*, 141–166.

Chagas, J. K. M., de Figueiredo, C. C., da Silva, J., & Paz-Ferreiro, J. (2021). The residual effect of sewage sludge biochar on soil availability and bioaccumulation of heavy metals: Evidence from a three-year field experiment. *Journal of Environmental Management, 279*, 111824.

Chen, Q. L., Fan, X.-T., Zhu, D., An, X.-L., Su, J.-Q., & Cui, L. (2018). Effect of biochar amendment on the alleviation of antibiotic resistance in soil and phyllosphere of Brassica chinensis L. *Soil Biology and Biochemistry, 119*, 74–82.

Chen, Y., Xu, Q., Sun, K., Han, L., Sun, H., Yang, Y., & Wang, Z. (2021). Effects of simulated diagenesis and mineral amendment on the structure, stability and imidacloprid sorption properties of biochars produced at varied temperatures. *Chemosphere, 282*, 131003.

Cheng, S., Chen, T., Xu, W., Huang, J., Jiang, S., & Yan, B. (2020). Application research of biochar for the remediation of soil heavy metals contamination: A review. *Molecules, 25*(14), 3167.

Cui, E.-P., Gao, F., Liu, Y., Fan, X.-Y., Li, Z.-Y., Du, Z.-J., Hu, C., & Neal, A. L. (2018). Amendment soil with biochar to control antibiotic resistance genes under unconventional water resources irrigation: Proceed with caution. *Environmental Pollution, 240*, 475–484.

Cycoń, M., Mrozik, A., & Piotrowska-Seget, Z. (2019). Antibiotics in the soil environment—Degradation and their impact on microbial activity and diversity. *Frontiers in Microbiology, 10*, 338.

Deolikar, R., & Patil, R. (2022). Recent advances in pesticides removal using agroindustry based biochar. *Development in Wastewater Treatment Research and Processes*, 265–290.

Gámiz, B., Velarde, P., Spokas, K. A., Celis, R., & Cox, L. (2019). Changes in sorption and bioavailability of herbicides in soil amended with fresh and aged biochar. *Geoderma, 337*, 341–349.

Gill, H. K., & Garg, H. (2014). Pesticide: Environmental impacts and management strategies. *Pesticides-Toxic Aspects, 8*(187), 10–5772.

Guo, M., Song, W., & Tian, J. (2020). Biochar-facilitated soil remediation: Mechanisms and efficacy variations. *Frontiers in Environmental Science, 183*.

He, L., Zhong, H., Liu, G., Dai, Z., Brookes, P. C., & Xu, J. (2019). Remediation of heavy metal contaminated soils by biochar: Mechanisms, potential risks and applications in China. *Environmental Pollution, 252*, 846–855.

Hernando-Amado, S., Coque, T. M., Baquero, F., & Martínez, J. L. (2019). Defining and combating antibiotic resistance from one health and Global Health perspectives. *Nature Microbiology, 4*(9), 1432–1442.

Hossain, M. Z., Bahar, M. M., Sarkar, B., Donne, S. W., Ok, Y. S., Palansooriya, K. N., Kirkham, M. B., Chowdhury, S., & Bolan, N. (2020). Biochar and its importance on nutrient dynamics in soil and plant. *Biochar, 2*, 379–420.

Issaka, E., Fapohunda, F. O., Amu-Darko, J. N. O., Yeboah, L., Yakubu, S., Varjani, S., Ali, N., & Bilal, M. (2022). Biochar-based composites for remediation of polluted wastewater and soil environments: Challenges and prospects. *Chemosphere, 297*, 134163.

Kołtowski, M., Hilber, I., Bucheli, T. D., & Oleszczuk, P. (2016). Effect of steam activated biochar application to industrially contaminated soils on bioavailability of polycyclic aromatic hydrocarbons and ecotoxicity of soils. *Science of the Total Environment, 566*, 1023–1031.

Krasucka, P., Pan, B., Ok, Y. S., Mohan, D., Sarkar, B., & Oleszczuk, P. (2021). Engineered biochar–A sustainable solution for the removal of antibiotics from water. *Chemical Engineering Journal, 405*, 126926.

Li, G., Khan, S., Ibrahim, M., Sun, T.-R., Tang, J.-F., Cotner, J. B., & Xu, Y.-Y. (2018). Biochars induced modification of dissolved organic matter (DOM) in soil and its impact on mobility and bioaccumulation of arsenic and cadmium. *Journal of Hazardous Materials, 348*, 100–108.

Lian, F., Yu, W., Zhou, Q., Gu, S., Wang, Z., & Xing, B. (2020). Size matters: Nano-biochar triggers decomposition and transformation inhibition of antibiotic resistance genes in aqueous environments. *Environmental Science & Technology, 54*(14), 8821–8829.

Liu, N., Charrua, A. B., Weng, C.-H., Yuan, X., & Ding, F. (2015). Characterization of biochars derived from agriculture wastes and their adsorptive removal of atrazine from aqueous solution: A comparative study. *Bioresource Technology, 198*, 55–62.

Mirsal, I. A. (2008). *Soil pollution* (pp. 117–136). Springer.

Morillo, E., & Villaverde, J. (2017). Advanced technologies for the remediation of pesticide--contaminated soils. *Science of the Total Environment, 586*, 576–597.

Murtaza, G., Ditta, A., Ullah, N., Usman, M., & Ahmed, Z. (2021). Biochar for the management of nutrient impoverished and metal contaminated soils: Preparation, applications, and prospects. *Journal of Soil Science and Plant Nutrition, 21*, 2191–2213.

Nyiramigisha, P. (2021). Harmful impacts of heavy metal contamination in the soil and crops grown around dumpsites. *Reviews in Agricultural Science, 9*, 271–282.

Pan, L., & Min. (2020). Biochar adsorption of antibiotics and its implications to remediation of contaminated soil. *Water, Air, & Soil Pollution, 231*(5), 221.

Quirantes, M., Nogales, R., & Romero, E. (2017). Sorption potential of different biomass fly ashes for the removal of Diuron and 3, 4-dichloroaniline from water. *Journal of Hazardous Materials, 331*, 300–308.

Sun, J., Pan, L., Tsang, D. C., Zhan, Y., Zhu, L., & Li, X. (2018). Organic contamination and remediation in the agricultural soils of China: A critical review. *Science of the Total Environment, 615*, 724–740.

Varjani, S., Kumar, G., & Rene, E. R. (2019). Developments in biochar application for pesticide remediation: Current knowledge and future research directions. *Journal of Environmental Management, 232*, 505–513.
Vithanage, M., Rajapaksha, A. U., Tang, X., Thiele-Bruhn, S., Kim, K. H., Lee, S.-E., & Ok, Y. S. (2014). Sorption and transport of sulfamethazine in agricultural soils amended with invasive-plant-derived biochar. *Journal of Environmental Management, 141*, 95–103.
Wang, J., & Shizong. (2019). Preparation, modification and environmental application of biochar: A review. *Journal of Cleaner Production, 227*, 1002–1022.
Wang, H., Lin, K., Hou, Z., Richardson, B., & Gan, J. (2010). Sorption of the herbicide terbuthylazine in two New Zealand forest soils amended with biosolids and biochars. *Journal of Soils and Sediments, 10*, 283–289.
Wang, Y., Liu, Y., Zhan, W., Zheng, K., Wang, J., Zhang, C., & Chen, R. (2020). Stabilization of heavy metal-contaminated soils by biochar: Challenges and recommendations. *Science of the Total Environment, 729*, 139060.
Xie, T., Reddy, K. R., Wang, C., Yargicoglu, E., & Spokas, K. (2015). Characteristics and applications of biochar for environmental remediation: A review. *Critical Reviews in Environmental Science and Technology, 45*(9), 939–969.
Yadav, R., Kumar, A., Tokas, D., & Singh, A. N. (2021). Soil contamination by polycyclic aromatic hydrocarbons in the agroecosystems. *Sustainable Agriculture Reviews 50: Emerging Contaminants in Agriculture*, 211–234.
Zeng, Z. W., Tian, S. R., Liu, Y. G., Tan, X. F., Zeng, G. M., Jiang, L. H., Yin, Z. H., Liu, N., Liu, S. B., & Li, J. (2018). Comparative study of rice husk biochars for aqueous antibiotics removal. *Journal of Chemical Technology & Biotechnology, 93*(4), 1075–1084.
Zhang, P., Ren, C., Sun, H., & Min, L. (2018). Sorption, desorption and degradation of neonicotinoids in four agricultural soils and their effects on soil microorganisms. *Science of the Total Environment, 615*, 59–69.

Chapter 6
Isotopic Tools for Tracing Water Pollutants

Zainab Tahir, Qaiser Farid Khan, Umar Hayyat, Faiza Sharif, Naba Azhar, Zaib un Nisa, Muhammad Farhan, Rizwan Rasheed, Gul Zareen Ghafoor, Muhammad Afzaal, Hamid Majeed, and Hafiz Arbab Sakandar

6.1 Introduction

Environmental isotopes include both naturally occurring and artificially introduced radioactive and stable isotopes. Environmental isotopes, especially those of light elements (hydrogen, carbon, nitrogen, oxygen, sulfur, and chlorine), are being used in the study of global element cycles, hydrology, food and drink authentication, medical research, and pollution monitoring due to their specificity and reliability.

Z. Tahir · U. Hayyat · F. Sharif · M. Farhan · R. Rasheed · G. Z. Ghafoor · M. Afzaal
Sustainable Development Study Centre, Government College University, Lahore, Pakistan
e-mail: umerhayyat@gcu.edu.pk; faizasharif@gcu.edu.pk; dr.muhammadfarhan@gcu.edu.pk; rizwanrasheed@gcu.edu.pk; zareen.sdsc@gcu.edu.pk; dr.afzaal@gcu.edu.pk

Q. F. Khan (✉)
Sustainable Development Study Centre, Government College University, Lahore, Pakistan

Department of Microbiology, IIIB, Government College University, Lahore, Pakistan
e-mail: dr.qaiserfarid@gcu.edu.pk

N. Azhar
School of Health Sciences, University of Management and Technology, Lahore, Pakistan

Z. u. Nisa
Department of Zoology, Government College University, Faisalabad, Pakistan
e-mail: zaibunnisa@gcuf.edu.pk

H. Majeed
Department of Food Sciences, Cholistan University of Veterinary and animal sciences, Bahawalpur, Pakistan

H. A. Sakandar
Shenzhen Key Laboratory of Marine Microbiome, Institute of Advanced Study, Shenzhen University, Shenzhen, China

Department of Food Science & Engineering, College of Chemistry & Environmental Engineering, Shenzhen University, Shenzhen, China

H. Alshemmari et al. (eds.), *Contaminated Land and Water*, Emerging Contaminants and Associated Treatment Technologies, https://doi.org/10.1007/978-3-031-65129-8_6

Their use in retracing a molecule's history relies on the fact that each molecule has a unique isotopic composition due to its unique origin and isotopic behavior as it is subjected to environmental processes like isotopic fractionation in purely physical processes, heterogeneous chemical equilibria, and reaction kinetics (i.e., any process involved in the transport and fate of the contaminant in the environment, including evaporation, dissolution, volatilization, sorption, and degradation). Molecules with a lower isotope react more quickly than those with a higher isotope because breaking bonds carrying a heavier isotope requires more energy in an isotopically fractionating process. The relative differences in the binding strengths of isotopes in different components of an equilibrium reversible system are reflected in the equilibrium isotope effects, as heavier isotopes tend to concentrate in the component to which the element is more closely coupled (Viers et al., 2018).

The very source of life is often endangered due to damage inflicted by human activities and infrastructure, which is ironic given that water is the source of all life. Nuclear techniques can aid in identifying issues with water quality and providing guidance on how to keep water fit for its intended usage and clean up polluted water. Heavy metals, complex organic molecules like petroleum byproducts or medications, radioactive isotopes, and trace elements can all contaminate water supplies and cause problems. The solubility of these contaminants in water and their ability to mix with minute particles control how far and how fast they travel across aquatic systems (also known as "scavenging"). Consequently, certain contaminants are promptly bound, while others can travel large distances (Wolkersdorfer et al., 2020).

Water quality can be impacted by both agricultural and urban runoff, as well as drainage water from mining operations. In many areas where crops are grown, the fertilizers seep into the ground and pollute the aquifers. The International Atomic Energy Agency (IAEA) contributes to water safety by modeling the effects of a variety of nitrate inputs on water purification schedules. The links between nutrient loads, eutrophication (high nutrient levels), and the proliferation of harmful algal blooms in freshwater systems are being investigated (Gopinath et al., 2019).

In a number of nations, increased quantities of trace elements in groundwater have emerged as a severe hazard to human health. One such example is increased arsenic levels in the water supply of the soil. The International Atomic Energy Agency uses uranium isotopes 238 U and 235 U to determine whether uranium in water originates from natural sources or the nuclear fuel cycle. The International Atomic Energy Agency (IAEA) can also evaluate the radioactive contamination of surface water left exposed to the air in storage ponds. This data was more significant than ever in light of the Chernobyl and Fukushima disasters (Tasker et al., 2020).

Efforts are concentrated on improving the quality of vital water sources. There have been positive alterations. The IAEA employs nuclear and isotopic techniques to examine the distribution of pollutants throughout water systems and to assess water quality. Using stable isotopes and radiotracers, researchers can track the travel of contaminants from their source to the eventual destination of the water, whether it be a lake, a river, or the ocean. Member states profit from this since it allows them to take steps to lessen the likelihood of contamination.

Radioactive tracers are chemical compounds containing radioisotopes at one or more atoms (or radiotracers). Even though radiotracers are only one type of environmental tracers, they play a crucial role in detecting and analyzing contaminants due to the fact that even minute concentrations of a particular radioisotope can be easily detected and the disintegration of short-lived isotopes does not leave any trace in the environment. For risk assessment, hydrologists utilize radiotracers to monitor the movement and concentration of contaminants in groundwater.

In regions with high levels of sulfate or sulfide pollution at the source, sulfur isotopes (especially sulfur-34, expressed as d 34S difference per mil, compared to a standard reference sample) and oxygen-18 of sulfates (expressed as difference per mil, compared to a standard reference sample) could be utilized. Sulfur-32, sulfur-33, and sulfur-36 are additional sulfur isotopes but ones that are rarely employed in hydrology. Oxygen-18 and nitrogen-15 (d 15 N) could be utilized as reference isotopes in nitrogen-rich regions (Cheema et al., 2020).

The isotopic tools that can be utilized for analyzing water pollutants include the following:

1. Compound-specific isotope analysis (CSIA).
2. Inductively coupled plasma mass spectrometry (ICP-MS).
3. Isotope fingerprint.
4. Isotope ratio infrared spectroscopy.
5. Inductively coupled plasma mass spectrometry.
6. Laser-induced fluorescence spectroscopy.
7. Thermal ionization mass spectrometer.
8. X-ray fluorescence spectrometry.
9. Accelerator mass spectrometry.
10. Monitoring and Identification of Water Pollutants.
11. Methodology.

Analyzing environmental contaminants in a variety of environmental matrices (e.g., water, soil, and sediment) and biological samples is feasible using MS (e.g., urine, serum, and tissues). Researchers are creating cutting-edge methods, such mass spectrometry-based metabolomics, proteomics, and MS imaging, to better understand the toxicological pathways of environmental contaminants. New approaches for analyzing environmental contaminants using MS should be prioritized. This chapter focuses on recent advances in mass spectrometry-based metabolomics and proteomics for analyzing water, soil, sediment, and other environmental media to learn more about chemical toxicity, biotransformation of contaminants with emerging concerns, and the fate and transport of pollutants.

Scientists rely on optical, gas chromatographic and mass spectroscopic techniques for a long time to identify minute quantities of gases, solids, and liquids in the air and water. Laser spectroscopy has served as the foundation for a variety of exciting new sensing and analytical measurement techniques, which have been successfully implemented in a vast array of situations. In all academic fields, the use of laser spectroscopy tools has expanded considerably. This passion has led to the creation of various valuable technologies, including diode and fiber lasers used in the optical communications industry, diode-pumped solid-state lasers, etc. As a result of these developments and the declining cost and complexity of laser systems, spectroscopic sources are becoming more broadly accessible and user-friendly in both existing and growing industries: air and water quality monitoring; industrial and traffic emissions; atmospheric chemistry; chemical analysis and process control; medicine, such as cancer recognition; national security, such as explosives detection; remote sensing of vegetation; characterizing art; etc. This chapter will discuss novel methods for analyzing environmental media, mass spectrometry for trace analysis, metabolomics and proteomics for assessing chemical toxicity, mass spectrometry imaging, biotransformation of emerging contaminants, and the occurrence and fate of environmental contaminants using isotopic tools.

6.1.1 Compound-Specific Isotope Analysis (CSIA)

The study of compound-specific isotopes permits us to monitor and evaluate the eradication of environmental pollutants (CSIA). Compounds containing light isotopes in the reactive position dissolve more quickly than those containing heavy isotopes. By monitoring isotope ratios, it is able to track deterioration processes throughout time and place. Monitoring the changes in the isotopic signatures of two or more elements can offer a more accurate understanding of degradation and the proportional contribution of various reaction pathways. Compound-specific isotope analysis (CSIA) may give a new line of evidence for addressing pollutant degradation in environmental systems at various temporal and spatial scales, according to studies of recognized chemicals in contaminated locations.

While CSIA has been around for some time for chemicals with contaminated histories, compounds with characteristic polar micropollutant structures are only now beginning to foster. Researchers found that the isotope ratios of micropollutants changed over the course of their experiments. However, as molecules get bigger, the rate at which the ratio of carbon isotopes changes slows down, since the effects of the isotopes on the reacting link are "diluted" by the nonreactive sites. Many micropollutants have a larger molecular size than typical legacy contaminants (chlorinated C2 compounds, benzene, and toluene), which means that degradation-induced changes in carbon isotope ratios may not be easily resolved within the variability of commercial sources unless the source value is well-constrained. Fortunately, this obstacle can be avoided by the use of multielement isotope analysis. Micropollutants typically have more components than traditional pollutants. With only one to three atoms per element, the effects of heteroatom isotopes are not as greatly attenuated as they are for carbon. For instance, nitrogen isotope fractionation may indicate aerobic diclofenac decomposition when carbon isotope fractionation is low.

The two most important analytical challenges that must be answered to generalize CSIA approaches to micropollutants are (1) the existence of extremely low (sub--mg/L) concentrations of micropollutants and (2) their polarity. First, massive amounts of water need to be removed in order to enrich enough analyte (often a few milligrams for repeated injections). Because of their low volatility, analytes are often lost less frequently during solid phase extraction (SPE); however, matrix cut-off during purge and trap is hindered. Using on-column injection for maximum sensitivity, SPE concentrates nonvolatile matrix components with target compounds, enhancing the matrix impact. CSIA of the triazine-class pesticide atrazine and its transition product desethylatrazine can be performed at concentrations below 1 mg/L after extract cleaning by preparative HPLC. The degradation of the potent insecticide a-cypermethrin was detected in batch soil testing after an extraction and cleaning procedure, demonstrating that CSIA may trace micropollutants sprayed or deposited on soil. Thus, it is important to account for the possibility of small isotope fractionation in repeated sorption-desorption cycles in soils and sediments. Due to their polar nature, micropollutants require a derivatization step and/or a selective sorbent phase before GC separation can be performed.

CSIA was able to identify sources of hexachlorocyclohexane (HCH) and determine the presence of HCH biodegradation under steady-state settings. Enantiomer-specific enrichment factors are available, and there are established instrumental approaches that can be used to achieve this goal. The benefits of integrating enantiomer ratios, CSIA, and ESIA at a geologically complicated landfill were demonstrated by the detection of phenoxyacetic acid pesticide degradation (dichlorprop and its degradate 4-CPP) with regionally changing enantioselectivity. However, CSIA processes can be developed, tested, and evaluated in both steady-state and event settings using smaller, controlled environmental systems in addition to field investigations. For instance, CSIA made it possible to learn that chloroacetanilide pesticide degrades in miniature wetlands.

6.1.1.1 Advances in CSIA

The micropollutant CSIA may be used in future studies that cover a wider area and take into account a wider range of sources and seasonal fluctuations. For more than three decades, integrated approaches based on isotopes have helped shape nitrate-related water management policy. To evaluate the causes of water pollution and the methods via which nitrate has been retained and moved within a watershed, it is necessary to analyze nitrate isotope data in conjunction with physicochemical and hydrological tracer data. An all-encompassing sample plan taking into account likely sources, hydrology, and land use is thus required.

Micropollutant CSIA gives a one-of-a-kind chance to monitor the degradation of persistent organic pollutants over extended time periods since CSIA data is not dependent on mass and time balances. For instance, the CSIA may shed light on the persistence of banned herbicides like atrazine in groundwater long after their prohibition in Europe (Won et al., 2021).

Chemical formulas for BOCs typically include carbon, nitrogen, hydrogen, oxygen, and bromine. Differentiating between two or more stable isotopes of a given element is a common application of mass spectrometry. In order to minimize systematic isotopic bias and properly calibrate isotopic data, the ratio of stable isotopes has been standardized according to worldwide reference standards set by the International Atomic Energy Agency (IAEA, Vienna, Austria). Because isotope shifts are typically reported in parts per million, the final number will require a factor of 1000 to convert from the more common metric. In order to make reliable comparisons between laboratories, it is necessary to standardize the isotopic composition data against an internationally recognized reference. These methods include two-point and multipoint linear normalization, as well as single-point anchoring and linear shift between measured and true isotopic compositions of two certified reference standards (Huang et al., 2020).

The ways in which CSIA can be used to study chemical processes (mechanisms) in nature and to detect organic contaminants are discussed. Isotopes have a long and fruitful history of application in agricultural and environmental investigations, suggesting that they can be used to investigate many different types of accidental pesticide contamination. The CSIA method can disclose the transformation mechanisms of organic pollutants in the natural environment by analyzing the way and timing of chemical bond breaking generated by chemicals. In contrast to traditional methods of analyzing the isotope composition of carbon, nitrogen, etc. in total organic matter, CSIA uses the stable isotope values of components involved in chemical reactions to determine the origin of specific substances, such as the formation or dissolution of chemical bonds in organic contaminants (such as bulk isotope analysis). These studies are ongoing in an effort to trace back the contamination to its original source and to examine the spread of biological movements and how they relate to environmental behavior. Micropollutants are notoriously difficult to analyze due to their minute amounts and high polarity, but this approach has shown promising results in addressing these challenges and opening up new avenues of inquiry into the mechanisms at play in environmental systems (Kuntze et al., 2020).

6.1.2 Inductively Coupled Plasma-Mass Spectrometry (ICP-MS)

As a highly sensitive method for (ultra)trace element analysis, it saw its first commercial application in 1983. When compared to atomic absorption spectrometry (AAS) and inductively coupled plasma optical emission spectrometry (ICP-OES), ICP-main MS excel due to its improved multielement capabilities and enhanced detection capacity (ICP-OES). ICP-MS, like mass spectrometers, provides isotopic data on the analyte elements it detects. This could be used to broaden the usefulness of ICP in practical contexts when elemental data is inadequate. Standard ICP-MS instruments are well suited for the study of induced and relatively considerable natural changes in the isotopic composition of target elements, as they employ a quadrupole filter or a sector field mass spectrometer for mass analysis and a single ion detector (Fig. 6.1). However, the intensity of the component ions can be measured simultaneously using multi-collector (MC) MC-ICP-MS systems, which use

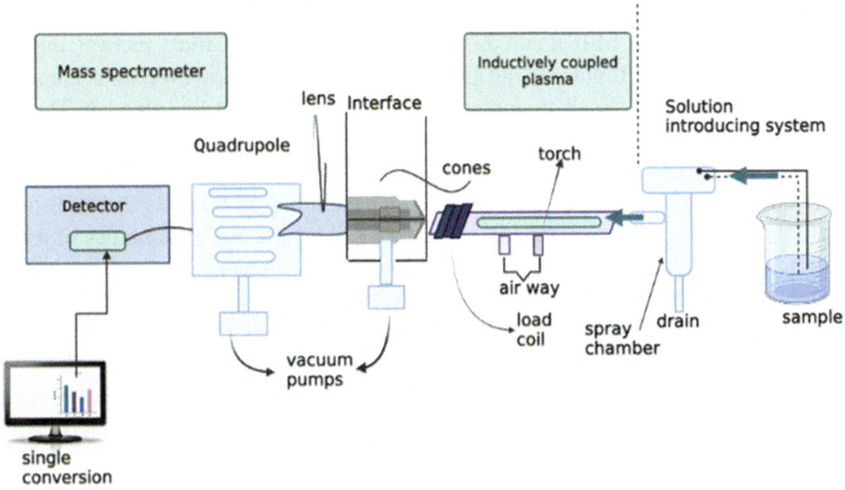

Fig. 6.1 Mechanism of inductively coupled plasma-mass spectrometry (ICP-MS)

a sector field mass spectrometer and an array of ion detectors. In this chapter, we will use real-world examples from the Department of Analytical Chemistry at the University of Ghent, demonstrating the versatility of isotopic analysis by ICP-MS through the use of both single- and multi-collector systems (Gaiss et al., 2019).

Multi-collector equipment (MC-ICP-MS) advances over the past three decades have made high-precision stable isotope analysis of nearly the entire periodic table possible, and stable isotope geochemistry is a rapidly expanding field. All the isotope ratios in the articles included in this thesis were measured using a Neptune PLUS MC-ICPMS from Thermo Scientific in Bremen, Germany. Isotope ratios can be measured with a single detector by detecting each isotope in turn. Sequential detection and the inconsistency of the plasma ion source (flicker noise) significantly degrade the reliability of the measurements.

In the late 1990s, the MC-ICP-MS mass spectrometer was developed. Several Faraday cups and ion counters are also included. In MC-ICP-MS, as opposed to TIMS, the substance is typically liquefied before being drawn into the plasma as an aerosol. Interference with argon-based MC-ICP-MS was seen at higher temperatures due to the presence of certain ion populations such as 40Ar12C, 40Ar16O, and 40Ar14N. High-resolution mass spectrometry is necessary for MC-ICP-MS to remove this ambiguity. The results of mass spectrometry of Cr isotopes by Schiller et al. (2014) using trifluoromethane (CHF3) as the carrier gas were consistent with those obtained using argon. Since the instrument bias for MC-ICP-MS is relatively constant throughout the analysis, the overall magnitude of mass bias is significantly greater than that of TIMS, and the sample matrix effect has a significant impact on the instrument mass bias, it is challenging to evaluate the accuracy of the isotope ratio. Evidence like this shows how important it is to account for potential biases in the larger population when conducting sample tests. Currently, the quality of samples for stable isotope identification is evaluated using the standard sample bracketing (SSB) and double-spike correction (DSC) protocols (DS). In order to be compatible with the SSB method, the chemical separation must recover the full amount of sample without causing any mass fractionation. However, most laboratories are finding it difficult to achieve complete recovery in this inquiry. The DS method, on the other hand, can provide precise elemental content at a high throughput without necessitating a second round of chemical separation. The DS strategy is the most popular option. To gain a better understanding of how DS and SSB mitigate the negative effects of bias in large populations, the 54Cr anomalies of astronomical objects including moon samples and iron meteorites are characterized using the SSB method. When attempting to correct for a meteorite's mass bias with the 50Cr-54Cr double-spike approach, it becomes challenging to account for 54Cr anomalies after adding 54Cr to the instrument (Sacchi et al., 2022).

Inductively coupled plasma mass spectrometry can assess trace element concentrations in biological fluids (ICP-MS). In the past decade, ICP-MS has steadily replaced earlier laboratory techniques such as atomic absorption and atomic emission. As this trend is likely to continue, it is essential for clinical scientists to get an understanding of the analytical features of ICP-MS, the potential for spectroscopic and non-spectroscopic interference, and the potential solutions for mitigating or eliminating these issues. ICP-MS has existed for decades, although many laboratories continue to employ obsolete techniques. ICP-MS is compared to the techniques of flame atomic emission (flame photometry), flame atomic absorption, atomic absorption in a graphite furnace, and atomic absorption in a cold vapor/hydride synthesis. From a laboratory's perspective, the advantage of ICP-MS is that it can simultaneously measure several components. Since only a tiny subset of elements can be absorbed by the atomic absorption lamp of a flame or graphite furnace, these techniques can only be used to measure one element at a time. ICP-MS is useful for rapidly processing a large quantity of laboratory samples due to its rapid analysis and minimal sample preparation requirements (Wei et al., 2020).

In addition to external calibration and standard additions, ICP-MS can also make use of the isotope dilution (ID) method of calibration thanks to its capacity to monitor isotope ratios. By combining a known amount of sample with an equal volume of spike (a standard solution of the analyte element itself but with a significantly different isotopic composition than the natural one), the isotope ratio of the target element may be determined in all three solutions. These readings can be used to determine the sample's concentration of the element of interest. Since an isotope ratio is far less sensitive to variations in instrument operating parameters or matrix composition, isotope identification is a particularly robust method. Institutions that produce reference materials are eager users of ID-MS because, when performed with the utmost care and control of all components impacting the final uncertainty, it gives exceptionally accurate and exact conclusions—useful even for everyday applications, particularly when a step of sample pretreatment is dedicated to separating out traces from the matrix. This is because the result is unaffected by analyte loss once isotopic equilibration has been reached (when the nuclides originating from the sample and the spike are mingled to the point that it is difficult to distinguish where a given nuclide originated). Recovery of the target element may be insufficient or even fluctuate during the separation process, but this poses no threat to the reliability of the results as long as enough of the element is preserved for a significant isotope ratio measurement to be made (Wassenaar et al., 2018).

6.1.3 Isotope Fingerprint

In fact, each water pollutant has a unique chemical fingerprint that can be used to find and tell them apart. The pattern of stable isotope ratios in a pollutant's different parts is called its "isotope fingerprint." Isotope ratio mass spectrometry is a method that scientists use to do this (IRMS). Environmental studies can learn a lot from IRMS because it can track where samples came from and find possible sources. Nitrates can come from the overuse of mineral nitrogen fertilizers and animal manure in farming, the release of nitrogen-containing municipal or industrial wastewater, the septic leachate in areas without sewers, the atmospheric deposition, and the breakdown of organic nitrogen in soil. Nitrate pollution is hard to deal with because there are so many possible sources of nitrate and because it can come from both current and past land uses (Kilic, 2019).

By looking at their unique mix of oxygen and nitrogen isotopes, scientists can figure out where nitrates come from. This lets an impact and cleanup strategy that is tailored to the source be put into place in a smart way. We can also now measure how much each source of nitrate pollution adds to the problem as a whole. This gives us a powerful tool for improving the quality of water around the world.

Even though strontium (Sr) isotope analysis is widely used to study how rocks break down and to track marine fluids, dust, and sediments, very few studies have tried to apply it to freshwater. Recent studies, however, have shown that Sr isotopes (87Sr/86Sr) combined with Sr concentration data are an effective way to fingerprint (identify) water sources and document aquifer linkages in a wide range of hydrologic settings and freshwater systems. As a geochemical tracer, one of the best things about Sr isotopes is that they aren't affected much by physical processes like rain, evaporation, and biological cycling. So, the 87Sr/86Sr ratios of both surface water and groundwater are the same as those of their parent minerals. This property can be used to directly identify potential Sr sources, whether they come from bedrock or are made by humans, like fertilizer, mine water, or industrial wastewater (Kilic et al., 2018).

Natural sample materials have an isotopic fingerprint that is unique to a place or process. This makes it possible to tell the difference between samples from different places (particulate matter in the air), plants (where organic matter comes from and how it moves), and mineralization processes (breakdown of material in nature). Carbon, nitrogen, sulfur, oxygen, and hydrogen isotope fingerprints can be used to trace these processes in a methodical way, showing where they came from and how the environment has changed from the recent past to now and in the future. Chemical fingerprints can be used to find and tell the difference between naturally occurring and man-made contaminants in the air, water, and soil. The stable isotopes of a

6.1.5 Laser-Induced Mass Spectrometry

Although ICP-SFMS has a number of amazing capabilities, not all ultra-trace elements may be identified in environmental/clinical matrices by direct analysis without analyte preconcentration and/or matrix separation. Although preconcentration and matrix-separation approaches enable the elimination of essential interferences, they do not significantly contribute to the MLOD. Because pollutants in anthropogenically damaged matrices (such as snow or freshwater samples) can be detected at low pg L1 values, the analytical methodologies described here are useful for pollution/exposure monitoring investigations (Biddau et al., 2019).

In order to measure the electronic energy levels of atoms and molecules, laser-induced fluorescence spectroscopy uses laser-induced electronic excitation (Whiteman et al., 2019). One possible outcome of an electron's fall in energy level is the release of a photon. To put it simply, this is the key idea of fluorescence. Many books and articles cover various theoretical, mathematical, and applied aspects of the LIF technique. As a result, we won't be covering them here. In this article, we will go through some of the fluorescence-based laser techniques that can be used for ecological studies (Wassenaar et al., 2018).

Laser-induced fluorescence (LIF) is a cost-effective and efficient environmental monitoring method (Fig. 6.2). It has been utilized in the research of the origins of dissolved organic matter (DOM) in aquatic ecosystems (e.g., the concentration of humic and fulvic acids, protein-like material, and phytoplankton-derived material). The amount of DOM that fluoresces when exposed to a particular wavelength of light can serve as a surrogate for the kind and concentration of DOM present. Blue,

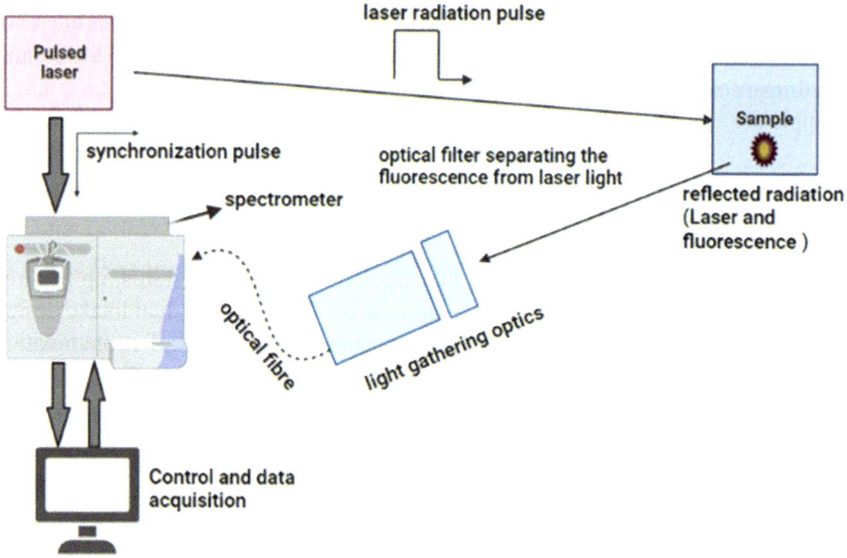

Fig. 6.2 Working of laser-induced fluorescence spectroscopy

violet, and ultraviolet light are particularly attractive to water dilute chemical compounds, such as humic and fulvic acids. Experiments on DOM contaminants in water have employed a vast array of excitation light sources (lamps, LEDs, and lasers). Several applications of induced fluorescence include the use of lasers and LEDs to sterilize water, the monitoring of water contamination, and the investigation of DOM composition in watersheds. Saraceno et al. examined the fluorescent proportion of colored DOM at 370 nm for excitation and 460 nm for emission (referred to as CDOM fluorescence or FDOM). Due to their versatility, diode lasers are widely used for detecting trace substances in the atmosphere. Diode lasers, like LEDs, are electrical components. There are numerous ways in which diode lasers distinguish themselves from the competition. These lasers are distinguished by their low input power, great spectrum purity, diminutive size, and minimal operating expenses. Since developments in semiconductor technology have made it possible for diode lasers to operate on tiny batteries, they are also less expensive than other types of lasers (Saribal, 2020).

There are a range of cutting-edge ways for conducting on-site water analysis. In two separate studies, LIBS was utilized to identify dissolved Ca, K, Li, Mn, and Na. Both calcium and sodium were subjected to pressures of up to 272 bar (4000 psi), but neither revealed evidence of being pressure-sensitive. We observed the expected rise in Mn emission with increasing pressure. The high salinity (NaCl) had little effect on the Mn and K signals, but the Ca signal increased (Wang & Zhang, 2019). Utilizing wooden substrates for water analysis is a novel and promising technology. Using LIBS, heavy metals such as Cr, Mn, Cu, Cd, and Pb were detected when the wood absorbed water. The LOD was determined to be between 0.029 and 0.59 mg/L, two to three orders of magnitude more sensitive than LIBS without a substrate. In some contemporary research, substrates, such as a metal plate perforated every 100 meters for LIBS analysis of soils, are utilized to provide a contained environment for the sample. According to the authors, keeping the sample encased in metal facilitates atomization and excitation. This method additionally reduces particle loss during irradiation (Deng et al., 2021).

6.1.6 Thermal Ionization Mass Spectrometer

Using what they had learned about the behavior of charged particles in electric and magnetic fields, scientists constructed thermal ionization mass spectrometry swiftly. The thermal ionization mass spectrometer (TIMS) is a "classic" instrument that remains at the forefront of various geoscience applications due to its capacity to attain unusually high isotope ratio precisions and/or extremely low detection limits (Fig. 6.3). Positives of the approach include its high sensitivity to elements that ionize well in thermal emission sources, the narrow energy range of the generated ions, and the low production rate of molecular backgrounds and multivalent ions (Barrat et al., 2020). Negative aspects of TIMS include the inability to effectively manage isotopic mass fractionation within an analysis or between studies, the requirement

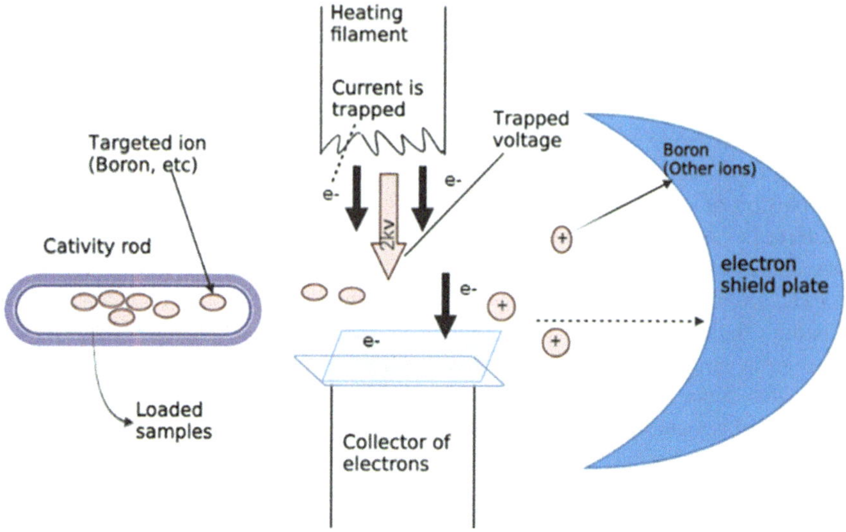

Fig. 6.3 Working of thermal ionization mass spectrometer

for lengthy sample preparation prior to analysis, and the very low sensitivity for elements that do not ionize quickly when heated. TIMS is the most sensitive or precise, or both, technique for isotope ratio analysis and concentration analysis by isotope dilution for a large number of geochemically/cosmochemically significant elements. Although TIMS faces competition from a variety of angles, its intrinsic benefits suggest that it will continue to be an essential instrument in geochemical/cosmochemical research for some time (Birch et al., 2021).

TIMS with a high number of ion collectors is frequently used to do isotope ratio measurements with an accuracy of 0.002%. While U interferences are less of an issue in TIMS, producing a source in the form of a thin filament requires lengthy sample preparation. 100 nanograms is the smallest amount of plutonium that may be safely sampled (109 g). After the filaments in the instrument's ion source are heated, the semivolatile analyte chemical is liberated as a gas. This vapor is ionized when it comes into touch with another hot electrode; plutonium's initial ionization potential is only 6.03 eV, making it an ideal analyte for this purpose. The equipment is pricey and effective operation requires training. It is possible to identify uranium and plutonium from a single particle with a DL in the picogram to femtogram range (1012–1015 g/sample). Numerous more radionuclides, including 41Ca, 241Am, 242Cm, 126Sn, 226Ra, 228Ra, and 99Tc, have been investigated using TIMS (Sankoh et al., 2021).

The accuracy of observed isotope ratios is limited by mass discrimination in the TIMS apparatus (such as the ion optical system or the ion detector) and mass fractionation effects (caused during the evaporation of sample, where the measured isotope ratio changes with time). Internal calibration techniques or isotopic standard reference materials with established isotope ratios for an element can be used to compensate for these limitations of TIMS isotope ratio measurements. For the most part, TIMS is used in the isotope dilution method, where highly enriched isotopic spikes are used to determine the isotopic composition of a sample (Sankoh et al., 2021).

Secondary ion mass spectrometry (SIMS) was shown to be capable of measuring the lead isotope ratios of individual particles by measuring lead-containing particles in three types of environmental certified reference materials (CRMs), including urban particulate matter, vehicle exhaust particulates, and coal fly ash. Electron probe microanalyzers equipped with wavelength dispersive X-ray spectrometers were used to identify and pinpoint the precise location of lead-containing particles. Particle-by-particle 207Pb/206Pb and 208Pb/206Pb isotope ratios on the carrier were determined using secondary ion mass spectrometry (SIMS). There was less than a 1.6% difference between the particles' relative standard deviations. Differences in lead isotope ratios were discovered between the three samples. Particles in a sample of exhaust particulate matter were sorted into two groups based on their lead isotope ratios (Li et al., 2019).

In contrast to more modern mass spectrometry methods, time-of-flight mass spectrometry (TIMS) uses a very simple ion source configuration. To create thermal ions, the sample is placed on a thin metallic filament that is heated resistively. Although TIMS ion sources appear straightforward, sample loading can be difficult and time-consuming due to a lack of knowledge about the ionization mechanisms. The sensitivity of TIMS measurements can be increased by a factor of 10, as shown in experiments (Rosca et al., 2018).

Mass spectrometric techniques are ideal for determining the precise composition of Pu isotopes in ultra-trace environmental contaminants, whether they are stable isotopes or radionuclides with lengthy half-lives. TIMS has the capability of identifying Pu at the femtogram level, making it one of the most sensitive methods for Pu investigation. While inductively coupled plasma-mass spectrometry (ICP-MS) has reached detection limits comparable to TIMS for other elements, it still lags behind for Pu analysis.

Since 238 U generates the isobaric 238UH+ ion in the plasma source, it confounds the detection of 239Pu and hence reduces the precision of the ICP-MS analysis of Pu. During operation, radioactive gaseous vapors are produced, calling for the use of ICP-MS and other forms of radioprotective gear. Despite these limitations, ICP-MS is widely utilized for applications such as discovering elemental impurities that reveal the origin and processing history of a chemical and performing bulk analyses of environmental contaminants. TIMS is typically utilized for highly sensitive and precise measurements. Despite the fact that both accelerator mass spectrometry (AMS) and resonance ionization mass spectrometry (RIMS) have been shown to attain sensitivity for Pu compared to TIMS, their practicality is constrained by the former's high cost and the latter's lack of available tandem accelerators (Millar et al., 2021).

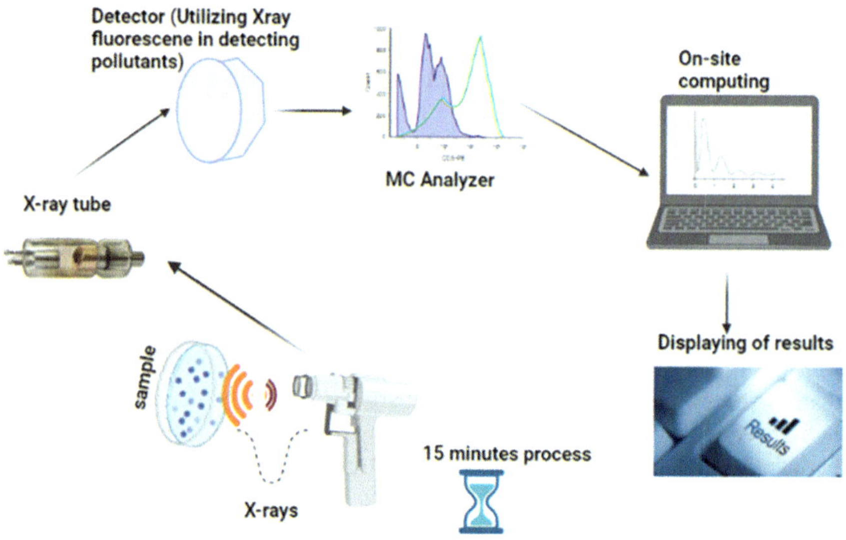

Fig. 6.4 Mechanism of X-ray fluorescence spectrometry

6.1.7 X-Ray Fluorescence Spectrometry

Using isotopically triggered energy dispersive X-ray fluorescence (XRF) spectroscopy (Fig. 6.4), the automated online quick real-time (5–15 min), simultaneous multicomponent (up to 20), and trace (1–10 ppb) evaluation of inorganic contaminants in reclaimed water was investigated. An examination of the anionic elements chromium-6, arsenic, and selenium is conducted. Due to the inherent insensitivity of XRF spectrometry to these elements, preconcentration is necessary prior to measurement (Pytlakowska et al., 2018). Several preconcentration techniques were evaluated, including direct and indirect evaporation, ion exchange membranes, selective and nonselective precipitation, and complexation (Zhang et al., 2018). It is demonstrated that XRF spectrometry is ideal for automated online quality assurance since it is a practical, nondestructive (samples may be kept for future use) analytical technique. In addition, the current statutory minimum detection levels may be attained without the use of high-powered X-ray tubes by combining an isotopically excited energy dispersive unit (50 mCi Cd-109 source) with a suitable preconcentration technique (Pytlakowska et al., 2020).

6.1.8 Accelerator Mass Spectrophotometer

Where the ions come from, since the isotopic ratios to be measured may be rather small, the ion production in an AMS system must meet specific criteria (down to 10–15 or less). In this aspect, cesium sputtering is the most common ion source for AMS. It is crucial that they have no effect on memory for this reason. Regularly occurring successive observations include a 300-second count of 129I particles at the detector and a 5-s reading of 127I current in the Faraday cup after the analyzing magnet (Pearson et al., 2018). In order to account for fluctuations in the 127I current generated by the sample, the 129I counts are divided by the average of the two 127I values obtained. Each measurement (cycle) is done a minimum of three times before moving on to the subsequent one. Following the testing of a standard sample and four unknown samples, a second standard sample is assessed. Each sample's measured 129I/127I ratio is normalized against the known nominal value of the standard's measured 129I/127I ratio (Carriere et al., 2020). Using the value of the relative ratio, the atomic number of 129I per liter of sample volume may be determined. In the supplemental materials, the analytical procedure is elaborately discussed (Quinto et al., 2019).

Using nuclear physics approaches such as element-specific energy loss in matter or the different deflection in a gas-filled magnet, AMS may reduce the quantity of stable isobaric background (such as 99Ru in the case of 99Tc and 135Ba in the case of 135Cs) in the final detection setup (Fig. 6.5). This becomes more challenging for nuclides with a lower relative difference in nuclear charge, as the isobaric interference of the isotopes of interest frequently arises from an adjacent element. Currently, 10Be–10B separation can be achieved with compact 300-kV tandem accelerators, but at higher masses, such as in the actinide mass area, even the largest AMS facilities with terminal voltages >10 MV struggle to achieve the needed isobar separation

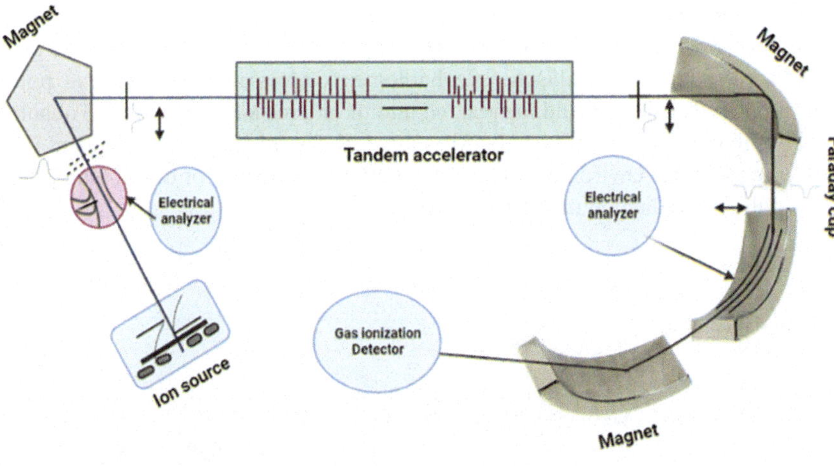

Fig. 6.5 Systemic working of accelerator mass spectrophotometer

of 106 or greater for isobaric systems (Tholkappian et al., 2018). To avoid this constraint and make nuclides such as long-lived fission products accessible at medium-sized AMS facilities with terminal voltages of 3 MV or less, VERA presented a technology that uses element-selective neutralization of anions by laser photodetachment. Five years ago, the ion-laser interaction mass spectrometry (ILIAMS) system at the VERA laboratory was successfully integrated with the preexisting AMS system (Proctor et al., 2020). In a radiofrequency quadrupole ion guide containing a buffer gas and overlapping with a collinear laser, the ion beam is slowed to energies in the eV range to allow for the necessary contact times between the ions and laser. Mukhopadhyay et al. (2020) have completed a comprehensive study of this approach and its results. For "typical" AMS isobaric systems, such as 36Cl-36S and 26AlO-26MgO, the ILIAMS technique provides isobar suppression of 1011–1012 (Seiwert et al., 2020).

6.2 Conclusion

In this chapter, we have explored several methods that could be utilized for tracing water pollutants that are effective in several purposes including sustaining the pollution sources with migratory measures. These techniques are effective in categorizing the pollutants with diverse data provided by each technique. By identifying and contrasting minute changes in isotope concentration or type, it is possible to change climate fluctuation effects with extreme accuracy. We can understand groundwater and surface water supply better by using environmental tracers and isotope techniques. In several nations throughout the world, the employment of isotopic technology has improved water resource management. To improve source identification accuracy, a range of isotope technologies (such as N, O, B, and Sr isotope) must be used. The recent application and refinement of the source analysis model have enabled objective assessment of the rate at which nitrate and other pollution sources contribute to water pollution. As a result, future research should focus on the combined application of several isotope technologies to precisely pinpoint nitrate pollution sources. The data should then be fed into the source analysis model to quantify the contribution rate of nitrate pollution sources in the freshwater environment. This will provide water environment managers with a foundation for developing prioritized pollution reduction methods.

References

Barbieri, M. (2019). Isotopes in hydrology and hydrogeology. *Water, 11*(2), 291.

Barrat, J. A., Bayon, G., Wang, X., Le Goff, S., Rouget, M. L., Gueguen, B., & Salem, D. B. (2020). A new chemical separation procedure for the determination of rare earth elements and yttrium abundances in carbonates by ICP-MS. *Talanta, 219*, 121244.

Biddau, R., Cidu, R., Da Pelo, S., Carletti, A., Ghiglieri, G., & Pittalis, D. (2019). Source and fate of nitrate in contaminated groundwater systems: Assessing spatial and temporal variations by hydrogeochemistry and multiple stable isotope tools. *Science of the Total Environment, 647*, 1121–1136.

Birch, Q. T., Potter, P. M., Pinto, P. X., Dionysiou, D. D., & Al-Abed, S. R. (2021). Isotope ratio mass spectrometry and spectroscopic techniques for microplastics characterization. *Talanta, 224*, 121743.

Carriere, S. D., Martin-StPaul, N. K., Cakpo, C. B., Patris, N., Gillon, M., Chalikakis, K., et al. (2020). Tree xylem water isotope analysis by isotope ratio mass spectrometry and laser spectrometry: A dataset to explore tree response to drought. *Data in Brief, 29*, 105349.

Cheema, A. I., Liu, G., Yousaf, B., Abbas, Q., & Zhou, H. (2020). A comprehensive review of biogeochemical distribution and fractionation of lead isotopes for source tracing in distinct interactive environmental compartments. *Science of the Total Environment, 719*, 135658.

Deng, X., Wu, Y., Liang, Y., Mao, L., Tan, Z., Feng, W., & Zhang, Y. (2021). Source apportionment of heavy metals in sediments of the urban rivers flowing into Haizhou Bay, eastern China: Using multivariate statistical analyses and Pb-Sr isotope fingerprints. *Environmental Science and Pollution Research, 28*(27), 36354–36366.

Gaiss, S., Amarasiriwardena, D., Alexander, D., & Wu, F. (2019). Tissue level distribution of toxic and essential elements during the germination stage of corn seeds (Zea mays, L.) using LA-ICP-MS. *Environmental Pollution, 252*, 657–665.

Gopinath, S., Srinivasamoorthy, K., Saravanan, K., & Prakash, R. (2019). Tracing groundwater salinization using geochemical and isotopic signature in southeastern coastal Tamil Nadu, India. *Chemosphere, 236*, 124305.

Huang, C., Zeng, Y., Luo, X., Ren, Z., Lu, Q., Tian, Y., et al. (2020). Tracing the sources and microbial degradation of PCBs in field sediments by a multiple-line-of-evidence approach including compound-specific stable isotope analysis. *Water Research, 182*, 115977.

Kilic, S. (2019). Survey of trace elements in bottled natural mineral waters using ICP-MS. *Environmental Monitoring and Assessment, 191*(7), 1–7.

Kilic, S., Cengiz, M. F., & Kilic, M. (2018). Monitoring of metallic contaminants in energy drinks using ICP-MS. *Environmental Monitoring and Assessment, 190*(4), 1–9.

Kuntze, K., Eisenmann, H., Richnow, H. H., & Fischer, A. (2020). Compound-specific stable isotope analysis (CSIA) for evaluating degradation of organic pollutants: An overview of field case studies. *Anaerobic Utilization of Hydrocarbons, Oils, and Lipids*, 323–360.

Li, C. F., Wu, H. Q., Chu, Z. Y., Wang, X. C., Li, Y. L., & Guo, J. H. (2019). Precise determination of radiogenic Sr and Nd isotopic ratios and Rb, Sr, Sm, Nd elemental concentrations in four coal ash and coal fly ash reference materials using isotope dilution thermal ionization mass spectrometry. *Microchemical Journal, 146*, 906–913.

Millar, C., Janzen, K., Nehemy, M. F., Koehler, G., Hervé-Fernández, P., & McDonnell, J. J. (2021). Organic contamination detection for isotopic analysis of water by laser spectroscopy. *Rapid Communications in Mass Spectrometry, 35*(15), e9118.

Mukhopadhyay, S., Chakraborty, S., Bhadoria, P. B. S., Li, B., & Weindorf, D. C. (2020). Assessment of heavy metal and soil organic carbon by portable X-ray fluorescence spectrometry and NixPro™ sensor in landfill soils of India. *Geoderma Regional, 20*, e00249.

Pearson, D., Weindorf, D. C., Chakraborty, S., Li, B., Koch, J., Van Deventer, P., et al. (2018). Analysis of metal-laden water via portable X-ray fluorescence spectrometry. *Journal of Hydrology, 561*, 267–276.

Proctor, G., Wang, H., Larson, S. L., Ballard, J. H., Knotek-Smith, H., Waggonor, C., et al. (2020). Rapid screening for uranium in soils using field-portable X-ray fluorescence spectrometer: A comparative study. *ACS Earth and Space Chemistry, 4*(2), 211–217.

Pytlakowska, K., Matussek, M., Hachuła, B., Pilch, M., Kornaus, K., Zubko, M., & Pisarski, W. A. (2018). Graphene oxide covalently modified with 2, 2′-iminodiacetic acid for preconcentration of Cr (III), Cu (II), Zn (II) and Pb (II) from water samples prior to their determination by energy dispersive X-ray fluorescence spectrometry. *Spectrochimica Acta Part B: Atomic Spectroscopy, 147*, 79–86.

Pytlakowska, K., Kocot, K., Pilch, M., & Zubko, M. (2020). Ultrasound-assisted dispersive micro-solid phase extraction using molybdenum disulfide supported on reduced graphene oxide for energy dispersive X-ray fluorescence spectrometric determination of chromium species in water. *Microchimica Acta, 187*(9), 1–10.

Quinto, F., Busser, C., Faestermann, T., Hain, K., Koll, D., Korschinek, G., et al. (2019). Ultratrace determination of 99Tc in small natural water samples by accelerator mass spectrometry with the gas-filled analyzing magnet system. *Analytical Chemistry, 91*(7), 4585–4591.

Rosca, C., Tomlinson, E. L., Geibert, W., McKenna, C. A., Babechuk, M. G., & Kamber, B. S. (2018). Trace element and Pb isotope fingerprinting of atmospheric pollution sources: A case study from the east coast of Ireland. *Applied Geochemistry, 96*, 302–326.

Sacchi, E., Cuoco, E., Oster, H., Paolucci, V., Tedesco, D., & Viaroli, S. (2022). Tracing groundwater circulation in a valuable mineral water basin with geochemical and isotopic tools: The case of FERRARELLE, Riardo basin, Southern Italy. *Environmental Geochemistry and Health, 44*(7), 1–28.

Sankoh, A. A., Derkyi, N. S. A., Frazer-williams, R. A., Laar, C., & Kamara, I. (2021). A review on the application of isotopic techniques to trace groundwater pollution sources within developing countries. *Water, 14*(1), 35.

Saribal, D. (2020). ICP-MS analysis of trace element concentrations in cow's milk samples from supermarkets in Istanbul, Turkey. *Biological Trace Element Research, 193*(1), 166–173.

Schiller, M., Van Kooten, E., Holst, J. C., Olsen, M. B., & Bizzarro, M. (2014). Precise measurement of chromium isotopes by MC-ICPMS. *Journal of Analytical Atomic Spectrometry, 29*, 1406–1416.

Seiwert, B., Klöckner, P., Wagner, S., & Reemtsma, T. (2020). Source-related smart suspect screening in the aqueous environment: Search for tire-derived persistent and mobile trace organic contaminants in surface waters. *Analytical and Bioanalytical Chemistry, 412*(20), 4909–4919.

Tasker, T. L., Warner, N. R., & Burgos, W. D. (2020). Geochemical and isotope analysis of produced water from the Utica/point pleasant shale, Appalachian Basin. *Environmental Science: Processes & Impacts, 22*(5), 1224–1232.

Tholkappian, M., Ravisankar, R., Chandrasekaran, A., Jebakumar, J. P. P., Kanagasabapathy, K. V., Prasad, M. V. R., & Satapathy, K. K. (2018). Assessing heavy metal toxicity in sediments of Chennai coast of Tamil Nadu using energy dispersive X-ray fluorescence spectroscopy (EDXRF) with statistical approach. *Toxicology Reports, 5*, 173–182.

Vaněk, A., Grösslová, Z., Mihaljevič, M., Ettler, V., Trubač, J., Chrastný, V., et al. (2018). Thallium isotopes in metallurgical wastes/contaminated soils: A novel tool to trace metal source and behavior. *Journal of Hazardous Materials, 343*, 78–85.

Viers, J., Grande, J. A., Zouiten, C., Freydier, R., Masbou, J., Valente, T., et al. (2018). Are cu isotopes a useful tool to trace metal sources and processes in acid mine drainage (AMD) context? *Chemosphere, 193*, 1071–1079.

Vystavna, Y., Schmidt, S. I., Diadin, D., Rossi, P. M., Vergeles, Y., Erostate, M., et al. (2019). Multi-tracing of recharge seasonality and contamination in groundwater: A tool for urban water resource management. *Water Research, 161*, 413–422.

Wang, H., & Zhang, Q. (2019). Research advances in identifying sulfate contamination sources of water environment by using stable isotopes. *International Journal of Environmental Research and Public Health, 16*(11), 1914.

Wassenaar, L. I., Terzer-Wassmuth, S., Douence, C., Araguas-Araguas, L., Aggarwal, P. K., & Coplen, T. B. (2018). Seeking excellence: An evaluation of 235 international laboratories conducting water isotope analyses by isotope-ratio and laser-absorption spectrometry. *Rapid Communications in Mass Spectrometry, 32*(5), 393–406.

Wei, X., Zhou, Y., Jiang, Y., Tsang, D. C., Zhang, C., Liu, J., et al. (2020). Health risks of metal (loid) s in maize (Zea mays L.) in an artisanal zinc smelting zone and source fingerprinting by lead isotope. *Science of the Total Environment, 742*, 140321.

Whiteman, J. P., Elliott Smith, E. A., Besser, A. C., & Newsome, S. D. (2019). A guide to using compound-specific stable isotope analysis to study the fates of molecules in organisms and ecosystems. *Diversity, 11*(1), 8.

Wolkersdorfer, C., Nordstrom, D. K., Beckie, R. D., Cicerone, D. S., Elliot, T., Edraki, M., et al. (2020). Guidance for the integrated use of hydrological, geochemical, and isotopic tools in mining operations. *Mine Water and the Environment, 39*(2), 204–228.

Won, E. J., Yun, H. Y., Lee, D. H., & Shin, K. H. (2021). Application of compound-specific isotope analysis in environmental forensic and strategic management avenue for pesticide residues. *Molecules, 26*(15), 4412.

Zhang, W., Tang, Y., & Xu, J. (2018). Online determination of boron isotope ratio in boron Trifluoride by infrared spectroscopy. *Applied Sciences, 8*(12), 2509.

Zhong, Q., Qi, J., Liu, J., Wang, J., Lin, K., Zhang, X., et al. (2022). Thallium isotopic compositions as tracers in environmental studies: A review. *Environment International, 162*, 107148.

Chapter 7
Nutrient Contamination in Water

Muhammad Farhan, Qaiser Farid Khan, Faiza Sharif, Maqsood Ahmad, Amina Kanwal, Zahid Ali Butt, Laila Shahzad, Shomaila Sikandar, Imran Afzal, Zaib Un Nisa, and Muhammad Ali

7.1 Introduction

One of the crucial resources that is needed to support human life is water. Freshwater is mostly obtained through rivers, lakes, and deep aquifers (Kamilya et al., 2022). But only around one-third of people worldwide have access to safe water. The increase in population leads to urbanization and industrialization, and the overuse of freshwater and groundwater has decreased both its quantity and quality (Oduor et al., 2023). In order to enhance water sustainability, activities including rainwater collection, desalination of saltwater, reuse of treated municipal, control agricultural

M. Farhan · F. Sharif · L. Shahzad · M. Ali
Sustainable Development Study Centre, Government College University, Lahore, Pakistan
e-mail: dr.muhammadfarhan@gcu.edu.pk; faizasharif@gcu.edu.pk; lailashahzad@gcu.edu.pk; dr.muhammadali@gcu.edu.pk

Q. F. Khan (✉)
Sustainable Development Study Centre, Government College University, Lahore, Pakistan

Department of Microbiology, IIIB, Government College University, Lahore, Pakistan
e-mail: dr.qaiserfarid@gcu.edu.pk

M. Ahmad
Department of Environmental Sciences, Baluchistan University of Information Technology, Engineering and Management Sciences, Quetta, Pakistan

A. Kanwal · Z. A. Butt
Department of Botany, Government College Women University, Sialkot, Pakistan

S. Sikandar · I. Afzal
Department of Biology, Lahore Garrison University, Lahore, Pakistan
e-mail: shomaila.sikandar@lgu.edu.pk; imranafzal@lgu.edu.pk

Z. U. Nisa
Department of Zoology, Government College University, Faisalabad, Pakistan
e-mail: zaibunnisa@gcuf.edu.pk

H. Alshemmari et al. (eds.), *Contaminated Land and Water*, Emerging Contaminants and Associated Treatment Technologies, https://doi.org/10.1007/978-3-031-65129-8_7

runoff, and industrial wastewater treatment are crucial (Kakade et al., 2021). Given the legal requirements and the principle of circular economy (CE), the idea of reusing treated wastewater and recovering resources from wastewater is a promising alternative for water management (Kamilya et al., 2022).

Reusing wastewater has become difficult due to the ongoing degradation of wastewater quality and strict regulations for its disposal or reuse (Kakade et al., 2021). In general, wastewater from domestic households, business facilities, agricultural runoff, fertilizers of plants, and industries contains considerable amounts of nitrogen and phosphorus, in addition to other organic and inorganic contaminants (Mohanavelu et al., 2022). Overfertilization introduces nitrogen and phosphorus into wastewater, which leads to groundwater contamination, nonpoint source pollution, and other environment-related problems (Kakade et al., 2021). High nitrogen and phosphorus levels in wastewater are also caused by a number of anthropogenic and natural activities (Mohanavelu et al., 2022). The amount of different nutrients including P and N used in the agricultural fields of different continents is variable. The highest nutrient contamination is observed in Asia and least in Africa. This is because of a number of factors including the population size, number of industries, environmental policies or laws, etc. (Kamilya et al., 2022).

In addition, groundwater contamination may result due to the leaching of excess of nutrients that are applied to the agricultural lands (Kamilya et al., 2022). Thyroid cancer, bladder/gastric cancer, and cardiovascular issues are just a few of the disorders that can develop from exposure to high concentrations of NO^{3-} and NO^{2-}. Contrarily, exposure to these substances may result in ionic disbalance in intercostal muscles, which increase respiratory failure and bronchial secretion (Oduor et al., 2023). Additionally, the release of GHGs and eutrophication is due to the high concentration of P and N in water bodies, causing a negative environmental impact (Bae & Seo, 2018).

Nutrient contamination is a growing environmental issue that can negatively affect the health of humans and wildlife, as well as the environment. Nutrients are substances that are necessary for life, such as P and N (Atieno et al., 2020). While these essential nutrients are required for human life to exist on Earth, an excess of them can damage water systems, killing fish and other aquatic organisms. It's important to understand how these contaminants enter our water systems in order to better protect our ecosystems from further damage (Cao et al., 2018). Nutrient contamination usually occurs when too much fertilizer or animal waste enters into surface water or groundwater sources. This typically happens when runoff from agricultural lands carries excessive amounts of nitrogen and phosphorus into nearby bodies of water (Chen et al., 2021). In addition, sewage treatment plants often release high levels of nitrogen into rivers and streams due to improper disposal practices (Bae & Seo, 2018).

Nutrient contamination in water is an increasing concern for many communities across the world. With rapid population growth and agricultural runoff, nutrient levels in bodies of water are rising, resulting in serious environmental damage (Atieno et al., 2020). Eutrophication can lead to oxygen depletion and an increase of toxic algal blooms that can be harmful to human health and aquatic ecosystems (Cao et al., 2018).

The economic effects of nutrient pollution on commercial and recreational fishing can be significant. According to the study, economies around the world are impacted by nutrient contamination in various ways and in different industries that are dependent on clean water (Chen et al., 2021). The US tourism industry lost almost $1 billion each year due to the recreational activities of the water bodies that are polluted due to nutrients. In Mississippi, tourists around the Gulf Coast account for $1.6 billion due to the less visitor's expenditures, 32% tax cuts, and 24,000 jobs loss (Chia et al., 2018).

7.2 Sources of Nutrient Contamination

Sources of nutrient contamination are varied and can include both natural and human-made activities.

7.2.1 Natural Sources of Nutrient Contamination

Natural sources of nutrient contamination include runoff from land that contains nitrogen and phosphorus. However, when too much enters waterways, it causes an overabundance of nutrients that can cause algal blooms to occur, leading to fish kills or other ecological damage (Chia et al., 2018). Due to snowfall and rain in towns and cities, the nutrients flow through hard surfaces, like roads, rooftops, and sidewalks, and transports pollutants into nearby water bodies. These nutrients come from the mixing of water currents as a result of weathering of soils and rocks (Chen et al., 2021). In coastal areas, the nutrients are released in the coastal water bodies though urban runoff and lack of treatment facilities for wastewater (Cao et al., 2018). Soil and wind erosions have also increased the amount of nutrients entering water sources (Atieno et al., 2020; Deng et al., 2020).

7.2.2 Anthropogenic Sources of Nutrient Contamination

The primary sources of contamination of nutrients come from agricultural runoff and wastewater discharge. Agricultural runoff is created when fertilizer used on farms leaches into the groundwater or is washed away by rainwater into rivers and streams (Deng et al., 2020). Crops require both P and N found in the form of animal dung and chemical fertilizers to grow. These nutrients can, however, be lost from farm fields when they are not fully absorbed by plants. This can have a detrimental effect on the quality of the air and water downstream (Gill & Malamud, 2017).

Wastewater discharge comes from household wastewater treatment plants which often release nitrogen-rich effluent into nearby bodies of water. Both septic and

which destroyed the habitat of the benthic organisms by submerged vegetation (Jiang et al., 2019). In addition, other biological effects, such as bottom water hypoxia, and the production of carbon dioxide due to the organic matter decomposition result in water acidification, accumulation of poisonous hydrogen sulfide, and modification of the biogeochemical processes (Dalu et al., 2019).

Excess P and N loss from the runoff of surface water from fertile soils depend on the time, rate of fertilizer, and quantity of water transported (Wurtsbaugh et al., 2019). Excess application of fertilizers and inadequate management techniques degrade the fertility of the soil (Dalu et al., 2019). Fewer amounts of P and N are lost to runoff water cultivated according to prudently directed and tested organic matter (Savic et al., 2022). Algal blooms are harmful and result in a variety of injuries to aquatic animals, such as poisoning by toxin secretion, clogging of fish gills, and localized anoxia, and result in the negative impact on fishing resources and the economy through the mass mortality of various aquatic lives (Wei et al., 2022). Additionally, increased levels of nitrogen and phosphorous stimulate plant growth which leads to overcrowding and competition between native species. This make some species unable to survive or reproduce successfully, ultimately leading to a decrease in biodiversity (Khan & Mohammad, 2014).

7.6 Nutrient Removal Strategies

Nutrient removal strategies are an important part of preserving water quality and preventing the growth of harmful algae in lakes, rivers, and other water sources. The two primary methods for nutrient removal involve biological and chemical treatment processes (Wu et al., 2011). Biological treatments involve introducing bacteria or microorganisms into the water that feed on nutrients, such as nitrogen and phosphorus (Jing et al., 2001). This process is often used in wastewater treatment plants to reduce the levels of these pollutants before effluent is discharged (Jing et al., 2001). Chemical treatments use chemicals such as alum or ferric chloride to bind with pollutants and make them settle out of the water column, so they can be more easily collected from the surface or removed from the bottom sediments (Jing et al., 2001). These processes help reduce nutrient levels without large-scale equipment/investments or extensive labor requirements. With the release of excessive nitrogen and phosphorus, wastewater treatment plants can contribute to the eutrophication of rivers and streams (Yadav et al., 2023). This is the world's most widespread water contamination crisis, causing serious algal blooms and oxygen depletion and leading to widespread erosion of freshwater ecosystems (Ma et al., 2023). Hence, it is clear that some specific effective measures must be developed to significantly lower the effects of nutrient pollution. A cost-effective, dependable, and environmentally friendly method of controlling point and nonpoint pollution can be an alternative (Ma et al., 2023).

7.6.1 Construction of Wetlands and Macrophytes for Nutrient Removal

Wetlands are one of the most significant links in the natural ecosystem and fulfill several critical functions. Wetlands are a key part of the system which preserve rebuilding nutrients and control floodwaters, thereby protecting areas (Ayres et al., 2023). Additionally, wetlands can help to accelerate the seeping out of contaminants from water flow (Ma et al., 2023). However, a growing number of reports on constructed wetlands experience water pollutants, with many wetlands not living up to their required ecological criteria, especially for nitrogen and phosphorous pollution (Yu et al., 2022). Macrophytes are considered to be one of the primary biological building blocks of constructed wetlands and play a critical role in wetland succession. They provide direct absorption of nitrogen and phosphorous, both of which are essential for the nourishment of wetland plants (Carrillo et al., 2022). Macrophytes are the intermediaries by facilitating bioremediation reactions. They increase the environmental heterogeneity in the rhizosphere and improve a large number of chemical and microbial reactions that contribute to the effectiveness of constructed wetlands (Carrillo et al., 2022). Several studies have reported that nutrient removal is significantly enhanced in many constructed wetlands with the use of macrophytes (Ma et al., 2022). Nitrogen and phosphorus removal processes in marine environments may vary with the plant species through the rate for oxidation of the wetland matrix, availability of labile carbon, transpiration by plant species, and changing seasons (Bedouh et al., 2022). Single-stage constructed wetlands cannot offer both aerobic and anaerobic conditions, which therefore limits the total nitrogen removal rate in these systems (Jing et al., 2001). Vertical flow constructed wetlands are effective at removing ammonia, but the process that occurs is not seen as effective (Ma et al., 2022).

7.6.2 Use of Biochars

Biochar has been widely used in soil to boost the agricultural productivity and as a low-cost adsorbent that recovers nutrients from wastewater and water. The prevalence of pharmaceutical substances in human excreta has dramatically increased over the years. These pharmaceuticals have nitrogenous compounds in them. Some of these contaminants might be picked up by biochar when it is used to recover nutrients. There is still knowledge gap in the biochar use for the recovery of nutrients without collecting various pollutants, despite extensive research and studies on the subject (Oduor et al., 2023). For contaminant adsorption on biochar, a variety of processes occur, such as electrostatic interactions, pore-filling processes, partitioning, hydrogen bonding, and hydrophobicity. A negative surface is needed for nutrient recovery. A negative surface charge on biochar makes it easier to adsorb ammonia for nutrient recovery, but the precipitation of metal phosphate is crucial for phosphate adsorption (Bol et al., 2018).

7.7 Conclusion

Nutrient contamination in the air, water, and soil poses harmful effects to the air, water, soil, and degradation of the natural sources. This also results to damage in human health, environment, and economy of the country. Now is the alarming situation to take steps to minimize and control these degradations; otherwise, it will lead to the unbearable loss which our ecosystem cannot sustain. Nutrient removal strategies should be devised and implemented by the state government to mitigate its negative effects and restore the natural resource. Wastewater treatment, use of microorganisms, constructed wet land, use of macrophytes, and use of biochar are some of the ways to remove nutrient contamination. Further research is needed on the new methods and techniques to reduce it. Government involvement as policymaker, grants provided, and monitoring are also very crucial to address the problem. Government departments or relevant R&D organizations should provide technical guidance to develop mechanisms for nitrogen and phosphorous removal. This will ultimately benefit the environment, human, and economy of the country.

References

Atieno, M., Herrmann, L., Nguyen, H. T., Phan, H. T., Nguyen, N. K., Srean, P., & Lesueur, D. (2020). Assessment of biofertilizer use for sustainable agriculture in the Great Mekong Region. *Journal of Environmental Management, 275*, 111300. http://dx.doi.org/10.1016/j.jenvman.2020.111300

Ayres, J., Awad, J., Walker, C., Page, D., van Leeuwen, J., & Beecham, S. (2023). Constructed floating wetlands for the treatment of surface waters and industrial wastewater. In *Regional perspectives of nature-based solutions for water: Benefits and challenges* (pp. 35–66). Springer International Publishing.

Bae, S., & Seo, D. (2018). Analysis and modeling of algal blooms in the Nakdong River, Korea. *Ecological Modelling, 372*, 53–63. http://dx.doi.org/10.1016/j.ecolmodel.2018.01.019

Bai, Z., Fan, X., Jin, X., Zhao, Z., Wu, Y., Oenema, O., & Ma, L. (2022). Relocate 10 billion livestock to reduce harmful nitrogen pollution exposure for 90% of China's population. *Nature Food, 3*(2), 152–160.

Bedouh, Y., Chihoub, D., Gherib, A., & Hamoud, F. (2022). Performance assessment of hybrid constructed wetlands planted with different macrophytes for industrial wastewater treatment. *International Journal of Environmental Studies, 23*, 1–15.

Board, O. S., & National Research Council. (2000). *Clean coastal waters: Understanding and reducing the effects of nutrient pollution*. National Academies Press.

Bol, R., Gruau, G., Mellander, P. E., Dupas, R., Bechmann, M., Skarbøvik, E., & Gascuel-Odoux, C. (2018). Challenges of reducing phosphorus-based water eutrophication in the agricultural landscapes of Northwest Europe. *Frontiers in Marine Science, 5*, 276.

Cao, L., Zhou, T., Li, Z., Wang, J., Tang, J., Ruan, R., & Liu, Y. (2018). Effect of combining adsorption-stripping treatment with acidification on the growth of Chlorella vulgaris and nutrient removal from swine wastewater. *Bioresource Technology, 263*, 10–16. http://dx.doi.org/10.1016/j.biortech.2018.04.094

Carrillo, V., Gómez, G., & Vidal, G. (2022). Phosphorus uptake by macrophyte plants in monocultures and polycultures in constructed wetlands for wastewater treatment. *Ecological Engineering, 182*, 106690.

Chen, X., Wang, K., Li, X., Qiao, Y., Dong, K., & Yang, L. (2021). Microcystis blooms aggravate the diurnal alternation of nitrification and nitrate reduction in the water column in Lake Taihu. *Science of the Total Environment, 767*, 144884. http://dx.doi.org/10.1016/j.scitotenv.2020.144884

Chen, M., Janssen, A. B., de Klein, J. J., Du, X., Lei, Q., Li, Y., & Liu, H. (2023). Comparing critical source areas for the sediment and nutrients of calibrated and uncalibrated models in a plateau watershed in Southwest China. *Journal of Environmental Management, 326*, 116712.

Chia, M. A., Jankowiak, J. G., Kramer, B. J., Goleski, J. A., Huang, I. S., Zimba, P. V., & Gobler, C. J. (2018). Succession and toxicity of Microcystis and Anabaena (Dolichospermum) blooms are controlled by nutrient-dependent allelopathic interactions. *Harmful Algae, 74*, 67–77. http://dx.doi.org/10.1016/j.hal.2018.03.002

Dalu, T., Wasserman, R. J., Magoro, M. L., Froneman, P. W., & Weyl, O. L. (2019). River nutrient water and sediment measurements inform on nutrient retention, with implications for eutrophication. *Science of the Total Environment, 684*, 296–302.

Deng, X., Li, D., Xue, C., Chen, B., Dong, J., Tetteh, P. A., & Gao, K. (2020). Cultivation of Chlorella sorokiniana using wastewaters from different processing units of the silk industry for enhancing biomass production and nutrient removal. *Journal of Chemical Technology & Biotechnology, 95*(1), 264–273. http://dx.doi.org/10.1002/jctb.6230

Du, H., Chen, Z., Mao, G., Chen, L., Crittenden, J., Li, R. Y. M., & Chai, L. (2019). Evaluation of eutrophication in freshwater lakes: A new non-equilibrium statistical approach. *Ecological Indicators, 102*, 686–692.

Gelli, R., Ridi, F., & Baglioni, P. (2019). The importance of being amorphous: Calcium and magnesium phosphates in the human body. *Advances in Colloid and Interface Science, 269*, 219–235.

Gill, J. C., & Malamud, B. D. (2017). Anthropogenic processes, natural hazards, and interactions in a multi-hazard framework. *Earth-Science Reviews, 166*, 246–269. http://dx.doi.org/10.1016/j.earscirev.2017.01.002

He, W., Wang, Q., Zhu, Y., Wang, K., Mao, J., Xue, X., & Shi, Y. (2021). Innovative technology of municipal wastewater treatment for rapid sludge sedimentation and enhancing pollutants removal with nano-material. *Bioresource Technology, 324*, 124675. http://dx.doi.org/10.1016/j.biortech.2021.124675

Jiang, Q., He, J., Wu, J., Hu, X., Ye, G., & Christakos, G. (2019). Assessing the severe eutrophication status and spatial trend in the coastal waters of Zhejiang province (China). *Limnology and Oceanography, 64*(1), 3–17.

Jing, S. R., Lin, Y. F., Lee, D. Y., & Wang, T. W. (2001). Nutrient removal from polluted river water by using constructed wetlands. *Bioresource Technology, 76*(2), 131–135.

Kakade, A., Salama, E. S., Han, H., Zheng, Y., Kulshrestha, S., Jalalah, M., & Li, X. (2021). World eutrophic pollution of lake and river: Biotreatment potential and future perspectives. *Environmental Technology & Innovation, 23*, 101604. http://dx.doi.org/10.1016/j.eti.2021.101604

Kamilya, T., Majumder, A., Yadav, M. K., Ayoob, S., Tripathy, S., & Gupta, A. K. (2022). Nutrient pollution and its remediation using constructed wetlands: Insights into removal and recovery mechanisms, modifications and sustainable aspects. *Journal of Environmental Chemical Engineering, 10*(3), 107444. https://doi.org/10.1016/j.jece.2022.107444

Khan, M. N., & Mohammad, F. (2014). Eutrophication: Challenges and solutions. In *Eutrophication: Causes, consequences and control* (pp. 1–15). Springer.

Lin, L., St Clair, S., Gamble, G. D., Crowther, C. A., Dixon, L., Bloomfield, F. H., & Harding, J. E. (2023). Nitrate contamination in drinking water and adverse reproductive and birth outcomes: A systematic review and meta-analysis. *Scientific Reports, 13*(1), 1–18.

Lintern, A., McPhillips, L., Winfrey, B., Duncan, J., & Grady, C. (2020). Best management practices for diffuse nutrient pollution: Wicked problems across urban and agricultural watersheds. *Environmental Science & Technology, 54*(15), 9159–9174.

Liu, L., Ouyang, W., Wang, Y., Lian, Z., Pan, J., Liu, H., & Niu, S. (2023). Paddy water managements for diffuse nitrogen and phosphorus pollution control in China: A comprehensive review and emerging prospects. *Agricultural Water Management, 277*, 108102.

Ma, Y., Huang, J., Han, T., Li, R., Cao, M., & Li, X. (2022). A comprehensive study on the exposure of nanoplastics to constructed wetland ecological systems: Macrophyte physiology and microbial enzymology, community composition and metabolic functions. *Chemical Engineering Journal, 434*, 134592.

Ma, S., Geng, H., Yan, C., Han, X., Xu, D., Zhao, F., & Wang, F. (2023). Nutrient dynamics and microbial community response in macrophyte-dominated lakes: Implications for improved restoration strategies. *Journal of Environmental Management, 325*, 116372.

Mohanavelu, A., Shrivastava, S., & Naganna, S. R. (2022). Streambed pollution: A comprehensive review of its sources, eco-hydro-geo-chemical impacts, assessment, and mitigation strategies. *Chemosphere, 300,* 134589. https://doi.org/10.1016/j.chemosphere.2022.134589

Ngatia, L., Grace, J. M., III, Moriasi, D., & Taylor, R. (2019). Nitrogen and phosphorus eutrophication in marine ecosystems. *Monitoring of marine pollution, 34*, 1–17.

Nguyen, T. T., Némery, J., Gratiot, N., Strady, E., Tran, V. Q., Nguyen, A. T., & Peyne, A. (2019). Nutrient dynamics and eutrophication assessment in the tropical river system of Saigon–Dongnai (southern Vietnam). *Science of the Total Environment, 653*, 370–383.

Oduor, N. A., Munga, C. N., Ong'anda, H. O., Botwe, P. K., & Moosdorf, N. (2023). Nutrients and harmful algal blooms in Kenya's coastal and marine waters: A review. *Ocean & Coastal Management, 233*, 106454. http://dx.doi.org/10.1016/j.ocecoaman.2022.106454

Rizzo, A., Sarti, C., Nardini, A., Conte, G., Masi, F., & Pistocchi, A. (2023). Nature-based solutions for nutrient pollution control in European agricultural regions: A literature review. *Ecological Engineering, 186*, 106772.

Sabo, R. D., Sullivan, B., Wu, C., Trentacoste, E., Zhang, Q., Shenk, G. W., & Linker, L. C. (2022). Major point and nonpoint sources of nutrient pollution to surface water have declined throughout the Chesapeake Bay watershed. *Environmental Research Communications, 4*(4), 045012.

Savic, R., Stajic, M., Blagojević, B., Bezdan, A., Vranesevic, M., Nikolić Jokanović, V., & Ondrasek, G. (2022). Nitrogen and phosphorus concentrations and their ratios as indicators of water quality and eutrophication of the hydro-system Danube–Tisza–Danube. *Agriculture, 12*(7), 935.

Wang, L., Shantz, A. A., Payet, J. P., Sharpton, T. J., Foster, A., Burkepile, D. E., & Vega Thurber, R. (2018). Corals and their microbiomes are differentially affected by exposure to elevated nutrients and a natural thermal anomaly. *Frontiers in Marine Science, 5*, 101–108.

Wei, Y., Cui, H., Hu, Q., Bai, Y., Qu, K., Sun, J., & Cui, Z. (2022). Eutrophication status assessment in the Laizhou Bay, Bohai Sea: Further evidence for the ecosystem degradation. *Marine Pollution Bulletin, 181*, 113867.

Wu, H., Zhang, J., Li, P., Zhang, J., Xie, H., & Zhang, B. (2011). Nutrient removal in constructed microcosm wetlands for treating polluted river water in northern China. *Ecological Engineering, 37*(4), 560–568.

Wurtsbaugh, W. A., Paerl, H. W., & Dodds, W. K. (2019). Nutrients, eutrophication, and harmful algal blooms along the freshwater-to-marine continuum. *Wiley Interdisciplinary Reviews: Water, 6*(5), e1373.

Xenopoulos, M. A., Barnes, R. T., Boodoo, K. S., Butman, D., Catalán, N., D'Amario, S. C., & Wilson, H. F. (2021). How humans alter dissolved organic matter composition in freshwater: Relevance for the Earth's biogeochemistry. *Biogeochemistry, 154*(2), 323–348.

Yadav, N., Ahn, H. J., Rane, N. R., Kurade, M. B., Li, X., Park, Y. K., & Jeon, B. H. (2023). Comprehensive study on removal of bisphenol-S and its metabolic fate using aquatic macrophytes. *Chemical Engineering Journal, 455*, 140967.

Yu, Q., Duan, X., Gu, Y., Li, J., Zhang, X., Chen, C., & Zhao, D. (2022). Increasing chemical oxygen demand and nitrogen removal efficiencies of surface-flow constructed wetlands in macrophyte-dominant seasons by adding artificial macrophytes. *Bioresource Technology, 348*, 126755.

Chapter 8
Biodegradation of POPs in Contaminated Land/Water

Qaiser Farid Khan, Mehvish Shabir, Gul Zareen Ghafoor, Faiza Sharif, Muhammad Khurshid, Rizwan Rasheed, Hafiz Arbab Sakandar, Naba Azhar, Umar Hayyat, Muhammad Tayyab Akhtar, Muhammad Afzaal, and Mohsin Gulzar Barq

8.1 Introduction

In the aftermath of the Second World War, scientists have identified a number of chemical pollutants that possess toxic properties, exhibit persistence in the environment, and are prone to long-range atmospheric transboundary migration and

---

Q. F. Khan (✉)
Sustainable Development Study Centre, Government College University, Lahore, Pakistan

Department of Microbiology, IIIB, Government College University, Lahore, Pakistan
e-mail: dr.qaiserfarid@gcu.edu.pk

M. Shabir · G. Z. Ghafoor · F. Sharif · R. Rasheed · U. Hayyat · M. Afzaal
Sustainable Development Study Centre, Government College University, Lahore, Pakistan
e-mail: zareen.sdsc@gcu.edu.pk; faizasharif@gcu.edu.pk; rizwanrasheed@gcu.edu.pk; umerhayyat@gcu.edu.pk; dr.afzaal@gcu.edu.pk

M. Khurshid
School of Biochemistry and Biotechnology, University of Punjab, Lahore, Pakistan
e-mail: khurshid.ibb@pu.edu.pk

H. A. Sakandar
Shenzhen Key Laboratory of Marine Microbiome Engineering, Institute of Advanced Study, Shenzhen University, Shenzhen, China

Department of Food Science & Engineering, College of Chemistry and Environmental Engineering, Shenzhen University, Shenzhen, China

N. Azhar
School of Health Sciences, University of Management and Technology, Lahore, Pakistan

M. T. Akhtar
Institute of Industrial Biotechnology, Government College University, Lahore, Pakistan

M. G. Barq
Department of Microbiology, University of Central Punjab, Lahore, Pakistan
e-mail: mohsin.barq@ucp.edu.pk

H. Alshemmari et al. (eds.), *Contaminated Land and Water*, Emerging Contaminants and Associated Treatment Technologies, https://doi.org/10.1007/978-3-031-65129-8_8

deposition. These contaminants, known as persistent organic pollutants (POPs), are predicted to have serious negative health effects on both human and animal populations, as well as marine biota in proximity to their sources of emission.

Over the past several decades, there has been a significant increase in land and water pollution caused by organic pollutants. POPs are a group of dangerous chemical compounds that originate from human activities related to production, use, and disposal (Alharbi et al., 2018). These pollutants pose a risk of adverse effects on human and wildlife health, as well as the environment in general. Exposure to POPs, even in low quantities, can lead to a variety of health issues, including reproductive disorders, increased cancer risk, autoimmune diseases, disruption of the endocrine system, impairments in neurobehavioral systems, increased risk of birth effects, and genotoxicity.

The need rundown of POP incorporates a large number of synthetic compounds with shifting physicochemical properties and toxicology, for example, pesticides like dichlorodiphenyltrichloroethane (DDT), hexachlorocyclohexanes (HCH), and hexachlorobenzenes (HCB), modern synthetic substances like polychlorinated biphenyls (PCBs), and inadvertent results of modern cycles like polychlorinated dibenzo-p-dioxins and dibenzofurans (PCDD/PCDF) and polycyclic sweet-smelling hydrocarbons (PAH).

The Arctic region has experienced particularly high levels of POPs, which has prompted environmental concerns among scientists. As a result, the Arctic Council was established in 1996, with a focus on addressing these concerns. The Council has been active in resolving environmental issues, with Finland assuming the chairmanship from 2000 to 2002 and conducting negotiations for the subsequent formation of the Stockholm Convention. Iceland subsequently took over the responsibility for implementing the convention from 2002 to 2004. However, in recent years, the Council's interest in POPs has decreased (Friedrich et al., 2016).

POPs are known to move toward colder regions, such as Alaska, where they sink as a result of lower temperatures. This results in a lack of breakdown of the settled contaminants and their persistence in the area for extended periods. These pollutants then move from air and water to soil and plants and ultimately to animals and humans. When pesticides were discovered in polar bear fat tissue in the late 1970s, it became evident that toxins from far-off sources could persist in the Arctic and have a significant impact on both wildlife and human health.

In response to this realization, the Stockholm Convention was adopted as a legal international agreement in 2001 (Wang et al., 2022). The convention's priority list of POPs, commonly referred to as the "dirty dozen," initially included 12 compounds. Over the past decade, the list has been updated to include compounds such as brominated flame retardants and perfluorinated alkyl substances. The convention continues to consider new compounds for inclusion on the list, such as hexabromocyclododecanes.

8.2 Categories of POPs

In the "dirty dozen," POPs include the following:

1. Heptachlor.
2. Hexachlorobenzene.
3. Mirex.
4. Toxaphene.
5. Polychlorinated biphenyls (PCBs).
6. Dioxins (polychlorinated dibenzo-p-dioxins and dibenzofurans).
7. Furans (polychlorinated dibenzo-p-dioxins and dibenzofurans.
8. Aldrin.
9. Chlordane.
10. DDT (dichlorodiphenyltrichloroethane).
11. Dieldrin.
12. Endrin.

The Stockholm (at UNEP, 2001) Show expects gatherings to go to lengths to kill or lessen the arrival of these 12 POPs. This can incorporate measures like limiting the utilization of specific POPs, advancing the utilization of choices, and carrying out guidelines to control the arrival of jumps into the climate. Furthermore, gatherings are expected to give an account of their advancement in carrying out the show, including any actions taken to diminish the utilization and arrival of POPs. The Stockholm Show additionally urges gatherings to make a move to address new POPs that have been distinguished yet not yet remembered for the show's rundown of need POPs (Ashraf et al., 2013). The show has a component for the assessment and posting of new POPs which is considered for the show's extensions (Fig. 8.1).

8.3 Sources of POPs

Once released into the environment, POPs can disperse through atmosphere and eventually settle in remote regions, such as the Arctic. Due to their persistence and resistance to degradation, POPs can gradually gather in the environment and in the tissues of living organisms, leading to potential negative effects on human and wildlife health. POPs can also be transported through the food chain, leading to a process known as biomagnification, where the concentration of POPs increases as it moves up the food chain (Ying et al., 2006).

The Stockholm Convention was adopted in 2001 and aimed to reduce and eliminate the production and use of POPs, as well as minimize their release into the environment. The convention's priority list of POPs, commonly referred to as the "dirty dozen," initially included 12 compounds. Over the past decade, the list has been upgraded to include compounds such as brominated flame retardants and perfluorinated alkyl substances. The convention continues to consider new compounds for inclusion on the list, like hexabromocyclododecanes.

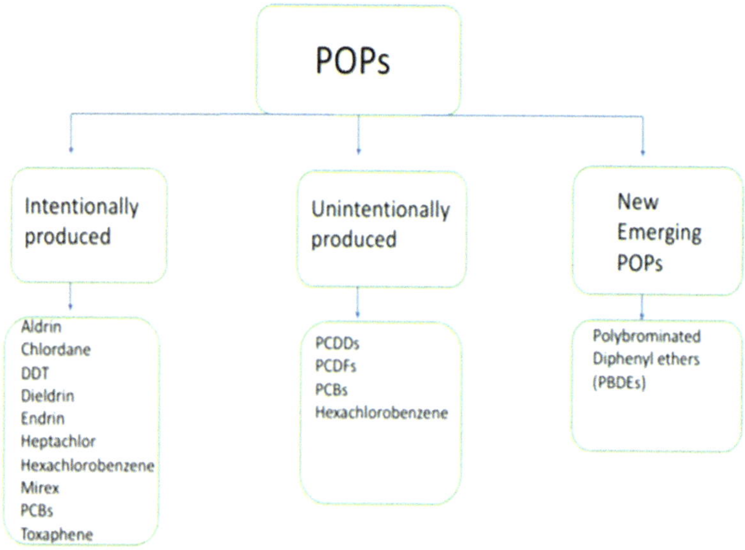

Fig. 8.1 Categorization of POPs

However, despite the efforts of the Stockholm Convention and other international agreements, POPs continue to be released into the environment and can still be found in the environment and in living organisms today. Therefore, it is essential to continue efforts to reduce the use and release of POPs and to monitor and assess the potential impacts of POPs on human and wildlife health. Some sources of POPs are shown in Fig. 8.2.

The major sources of POPs include:

1. The manufacturing sector.
2. Thermal processes.
3. Recycling processes.
4. Waste disposal and reservoirs.
5. Product application and use.

8.4 Effects of POPs

8.4.1 Health Effects

Openness to tireless natural contaminations (POPs) can bring about a scope of negative well-being impacts, including fatalities, constant sicknesses, and formative irregularities. A few POPs are known cancer-causing agents and might be connected to an expanded gamble of bosom malignant growth. Furthermore, many POPs are

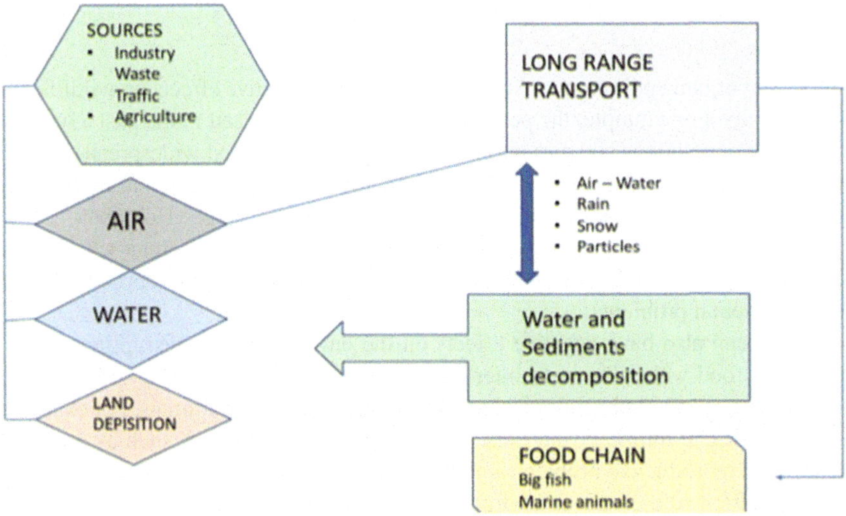

Fig. 8.2 Schematic depicting POPs in the environment and environmental processes during long-range atmosphere transport, bioaccumulation, and biomagnification

fit for disturbing the endocrine framework, which can influence the invulnerable framework, focal sensory system, and conceptive framework.

The principal ways that individuals and creatures are presented to POPs are through their eating regimen, either through the food they eat or while creating in the belly (Ritter et al., 2007). For people not presented to POPs unintentionally or at work, more than 90% of openness comes from creature food items as POPs bioaccumulate in fat tissues and move up the pecking order. Serum levels of POPs frequently increment with age and are normally higher in females than males.

The effects of POPs in adults can include:

Effects of adults	Symptoms	References
1. Reproductive system	2. Reduced fertility 3. Weakened immune system 4. Beginning of early puberty	El-Shahawi et al. (2010)
5. Endocrine disruption	Alters the functions of the hormonal system which may cause to many disorders 6. Pigmentation of nail 7. Liver problem 8. Acne and skin rashes	Damstra (2002)
9. Infant head circumference and gestational weight gain	10. Disability in growth of fetus 11. Decreased the weight of birth 12. Head and chest circumference	Dewan et al. (2013)

8.4.2 Environmental Effects

Persistent organic pollutants are also known to have negative effects on wildlife and biodiversity. For example, the pesticide DDT was widely used in the past to control mosquito populations and agricultural pests, but it also caused widespread declines in bird populations, including the bald eagle and the peregrine falcon, by thinning their eggshells and making them more susceptible to breakage (Holoubek, 2011). Additionally, several species of marine mammals, such as the beluga whale, have been found to have high levels of POPs in their tissues, leading to reproductive and developmental problems.

POPs can also have negative effects on the environment by disrupting ecosystems and food webs. They can interfere with the normal functioning of hormones, leading to a range of physiological and behavioral changes in animals. For example, some POPs can mimic or interfere with the action of hormones such as estrogen, leading to reproductive problems and other health issues.

In addition to their adverse effects on wildlife, POPs can also have significant economic impacts. For example, the decline in bird populations caused by DDT led to a decline in the number of birds available for hunting and a decrease in the value of hunting licenses. Similarly, the decline in fish populations due to POPs can lead to a decline in commercial and recreational fishing industries (Urbaniak, 2013).

Overall, POPs pose significant environmental and ecological risks, with potential impacts on human health and wildlife populations, as well as negative economic impacts. Therefore, it is crucial to continue efforts to reduce the use and release of POPs and to monitor and assess their potential impacts on the environment. Decomposition of PCDD/PCDF and PCBs is shown in Fig. 8.3.

8.4.2.1 Aquatic Life's Effects

In recent decades, the impact of persistent organic pollutants (POPs) on aquatic life has become increasingly apparent. Two types of POPs have been documented to have the greatest impact on fish populations: polychlorinated biphenyls (PCBs) and organochlorine pesticides (OCs). Fish can be exposed to POPs directly or indirectly through the ingestion of other organisms or sediments. These pollutants can accumulate in the tissues of fish, leading to an increase in concentration (biomagnification) as they move up the food chain, with predator fish having the highest concentrations (Authman et al., 2015).

Experimental studies have shown that exposure to multiple POPs can lead to an increase in the disruption and mortality of the endocrine system in fish, known as the "cocktail effect" (Khezri et al., 2017). POPs can also cause variations in reproductive tissues and affect thyroid hormone levels in adult fish. Additionally, research has shown that fish in Norwegian lakes are contaminated with high levels of POPs, leading to increased rates of tuberculosis and parasitic infections (Berg et al., 2016).

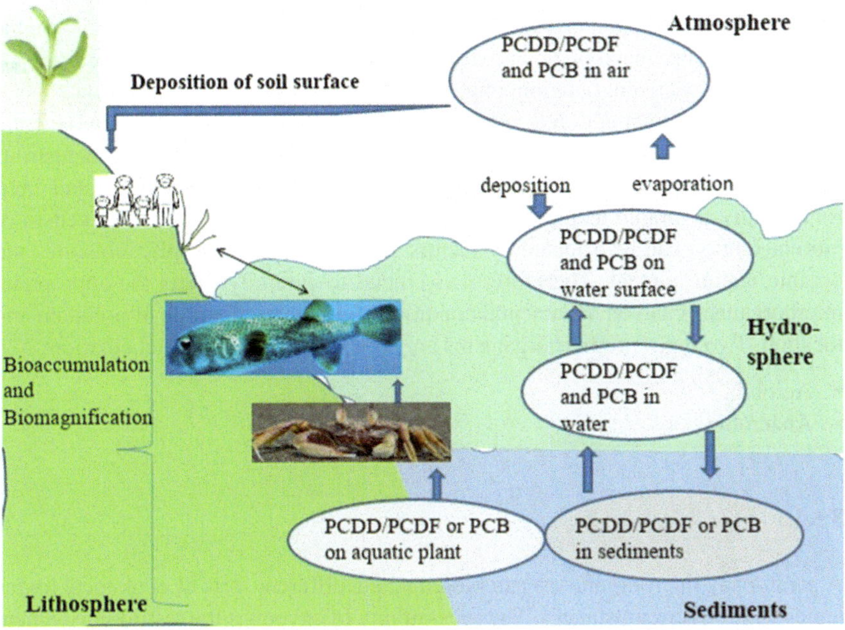

Fig. 8.3 Transport and circulation of PCDD/PCDF and PCBs in the environment

POPs can also affect the reproduction rate in adult fish by impairing maturation, fecundity, secondary sexual characteristics, and egg or larval size.

8.5 Remediation Technologies for Biodegradation of POPs in Contaminated Land/Water

8.5.1 Bioremediation by Microorganisms

Autochthonous microorganisms are naturally present in the environment, and inoculated microorganisms are introduced to the environment to perform specific functions. Autochthonous microorganisms have the advantage of being well adapted to the environment, while inoculated microorganisms have the advantage of being specifically chosen for their ability to degrade certain pollutants.

Bioremediation can be further divided into two categories: biostimulation and bioaugmentation. In order to promote the growth and activity of naturally occurring microorganisms, a process known as biostimulation includes adding nutrients to the environment. Bioaugmentation involves the addition of microorganisms that are specifically chosen for their ability to degrade specific pollutants.

One of the most popular types of bioremediation is phytoremediation, which uses plants to remove pollutants from the environment (Enerijiofi, 2021). Plants have the ability to take up, metabolize, and degrade pollutants through various mechanisms such as rhizosphere degradation, phytovolatilization, and phytostabilization.

Microbes are great option for remediation due to their biological system (Hussain et al., 2022). Overall, persistent organic pollutants (POPs) and other dangerous compounds can be removed from the environment using the economical and environmentally favorable technique of bioremediation. However, the effectiveness of bioremediation can be affected by factors such as temperature, pH, moisture, and organic matter content. Therefore, it is crucial to carefully select the appropriate microorganisms and environmental conditions for the specific type of pollutant and location. Two types of microorganisms are commonly used in bioremediation:

- Aerobic.
- Anaerobic.

8.5.1.1 Aerobic

A great many microorganisms can bioremediate different sorts of ecological toxins under vigorous circumstances. The accessibility of oxygen is the restricting element in the vigorous bioremediation process for the development of microorganisms. High-impact microorganisms like *Bacillus*, *Pseudomonas*, *Sphingomonas*, *Flavobacterium*, *Nocardia*, *Rhodococcus*, and *Mycobacterium* can separate complex natural mixtures. These microorganisms have been found to corrupt polyaromatic synthetic substances, alkane hydrocarbons, and pesticides by involving them as a wellspring of carbon and energy (Tegene & Tenkegna, 2020).

8.5.1.2 Anaerobic

A land- and/or water-capable microorganisms have the capacity to separate and change over foreign substances into less risky structures and have been applied in the remediation of polychlorinated biphenyls, chlorine compounds, and chlorinated solvents like trichloroethylene and chloroform (Tegene & Tenkegna, 2020). Microorganisms like *Pseudomonas*, *Aeromonas*, and sulfate-diminishing microscopic organisms have additionally been used in the anaerobic bioremediation process.

8.5.2 POP Degradation by Microbial Glycoconjugate

Besides, research has uncovered that microbial glycoconjugates assume a part in the upgraded corruption of persistent organic pollutants (POPs) from the climate. These glycoconjugates increment the bioavailability of these contaminations by lessening surface strain and making a dissolvable connection point. Nonetheless, there has been restricted accentuation on the capability of glycoconjugates in biodegradation through digestion.

8.5.2.1 Glycoconjugate Microbes Used in Biofilm Formation

Microbial glycoconjugate are used in biofilm formation for remediation of organic pollutants. Microbial community produces glycoconjugate which showed cumulative effect for degradation of organophosphorus pesticide (OPs), and many microorganisms are used in microbial "biofilm" which can degrade wide range of pollutants. Remediation technology which is based on biofilm is most effective, environment-friendly, and easily accessible for removing the pollutants from environment. Microbial film formation is also used as monitoring of heavy metal pollutants in water.

8.5.3 Biodegradation Pathways of POPs

POPs are resistant to the majority of environmental degradation processes, but there are some molecular changes that can occur in the environment that may not always result in molecules that are simpler and less dangerous. Some POP metabolites are just as complicated and hazardous as the parent chemicals. Microorganisms support the majority of the environmental POP degrading mechanisms. Some examples of POP degradation are represented below.

8.5.3.1 Degradation of Heptachlor

It is known that heptachlor can be dechlorinated by oxidation in the soil, forming 1-hydroxychlordene, and through the action of soil microbes, heptachlor epoxide, which has a half-life of up to 112 days as shown in Fig. 8.4 (Hayashi et al., 2013).

8.5.3.2 Degradation of DDT

Enterobacter aerogene bacteria can break down p,p'-DDT in soil to reduce dehydrochlorinated chemicals, DDE, and DDD, as well as oxidized derivatives that eventually become p,p'-dichlorobenzophenone in the presence of UV radiation and/or an iron catalyst (Zacharia & Croatia, 2011) as shown in Fig. 8.5.

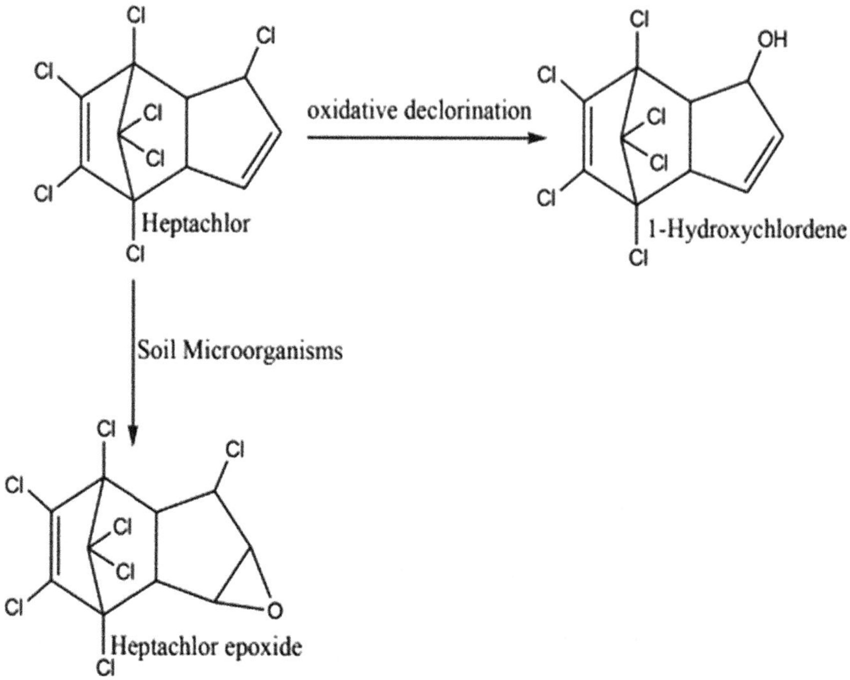

Fig. 8.4 Degradation of heptachlor

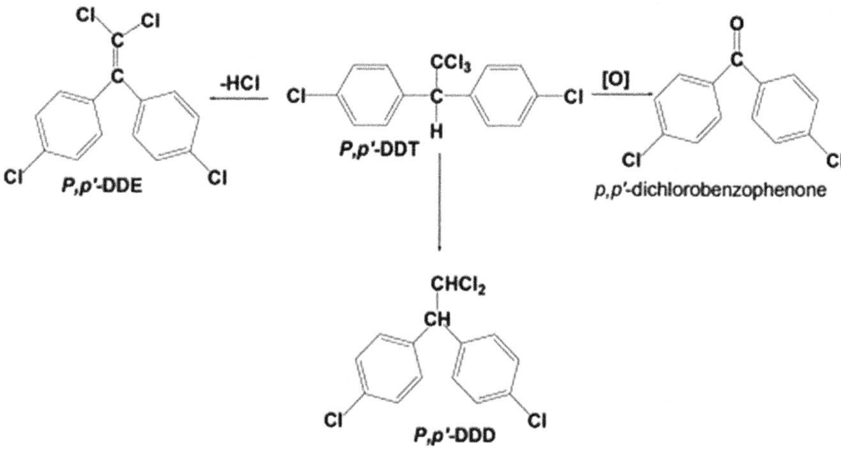

Fig. 8.5 Degradation of p'p-DDE in environment

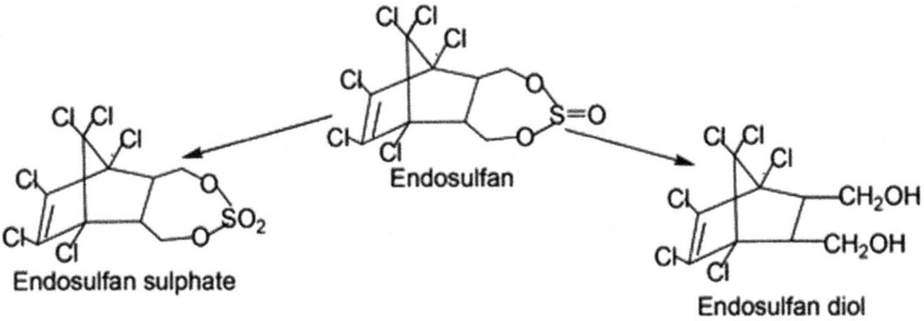

Fig. 8.6 Degradation of endosulfan

8.5.3.3 Degradation of Endosulfan

The alpha- and beta-isomers of endosulfan make up 70:30 of the technical grade's two isomers. Endosulfan is photolytically broken down in the environment to produce endosulfan sulfate in soil and endosulfan diol in aquatic environments (Fig. 8.6). It is equally harmful to humans as the parent chemical is endosulfan sulfate (Zacharia & Croatia, 2011).

8.5.3.4 Degradation of PCBs

PCBs are too chemically stable to be employed as energy sources. In anaerobic respiration, PCBs can be employed as electron acceptors, which is also the first step in the treatment route, reductive dechlorination, because of the chlorination, to store energy. Once the PCBs have undergone a specific amount of dechlorination, typically less than five chlorines are present in the structure, and one aromatic ring is chloride-free. In an aerobic environment, they can be broken down through the biphenyl (Bp) degradation pathway (BP pathway) into accessible carbon or CO2 (Fig. 8.7). Biphenyl is converted to TCA cycle intermediates (pyruvate and Acyl-CoA), benzoate, and other compounds via the BP pathway, which uses a number of enzymes (BphA, BphB, BphC, BphD, BphE, BphF, and BphG) (Kearney & Kaufaman, 1973).

8.5.4 *Fungi-Based*

The utilization of arbuscular mycorrhizal growths (AMF) in phytoremediation of tireless natural toxins (POPs) has been the subject of exploration, with fluctuating outcomes. A few examinations have demonstrated that AMF species from tainted regions might be more viable in further developing plant development and remediating hydrocarbons contrasted with economically accessible inoculants. Nonetheless,

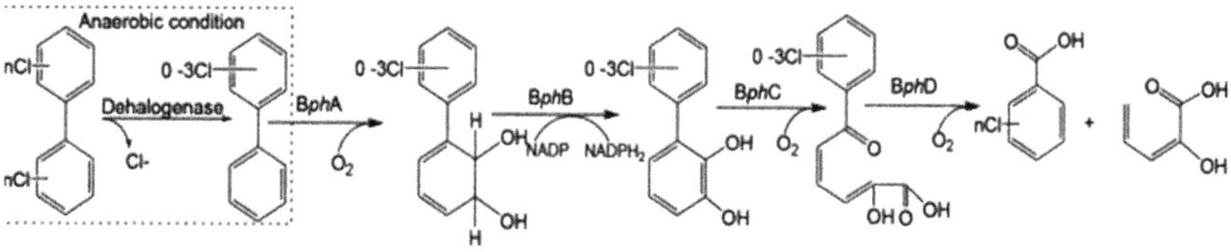

Fig. 8.7 Degradation of PCBs in the environment

the commitment of AMF colonization to POPs dispersal stays disputable, as it might differ relying upon variables like the sort of AMF, plant species, and centralization of POPs (Wang et al., 2011). Field tests have shown superior dispersal of polycyclic aromatic hydrocarbons (PAHs) and polychlorinated biphenyls (PCBs) in soil colonized by mycorrhizal plants (Sieverding et al., 2014). Different cycles, like biodegradation and volatilization, might be engaged with the decrease of POPs in mycorrhizal soil.

8.5.4.1 Direct Dispersion of POP AMF

The correlation between the degradation of persistent organic pollutants (POPs) and extracellular ligninolytic enzymes, such as peroxidases and laccases, has been observed in ligninolytic fungi. Studies have also demonstrated the positive impact of symbiotic associations on the dissipation of anthracene in the absence of other microorganisms. Additionally, genome sequencing of the fungus *R. irregularis* has revealed the presence of genes potentially involved in the degradation of fluorene, naphthalene, and anthracene, as well as those involved in the cytochrome P450 metabolism of xenobiotics.

It is important to note that while the use of arbuscular mycorrhizal fungi (AMF) and ligninolytic fungi has shown promise in the phytoremediation of POPs, more research is needed to fully understand the mechanisms and effectiveness of these organisms in the degradation of POPs. Additionally, further studies are needed to determine the optimal conditions for the growth and colonization of these fungi, as well as the appropriate combination of fungi and plant species for specific types of POPs. Overall, the use of AMF and ligninolytic fungi in phytoremediation is a promising approach, but more research is needed to fully understand their potential and limitations.

8.5.4.2 Bioremediation of Organic Agrochemicals (OAC) in Fungi-Based Soil Systems

Multiple investigations have highlighted the significance of fungi in the degradation and alteration of persistent agrochemicals. The enzymes produced by ligninolytic fungi have been demonstrated to possess the ability to break down resistant compounds (Čvančarova et al., 2015). Additionally, fungi have been found to engage in both physical and chemical interactions with organic chemicals, resulting in structural modifications or complete degradation of the target molecules.

8.5.4.3 Enhanced AMF-Assisted Phytoremediation of POP-Contaminated Soil Using New Techniques

It has been proposed that the dispersion of persistent organic pollutants (POPs) and the efficacy of arbuscular mycorrhizal fungi (AMF) phytoremediation may be improved through the implementation of strategies aimed at enhancing the solubility and biodegradability of POPs:

1. One potential approach that has been suggested is the utilization of surfactants, which have the potential to increase the desorption and bioavailability of POPs, as well as to promote the biodegradation of aged POPs in soil. However, it is worth noting that certain surfactants, such as Tween 80 nonionic surfactants, have been found to possess biological toxicity, which may negatively impact environmental safety.
2. As an alternative, biosurfactants such as rhamnolipid and surfactin, which are naturally harmless and may try and advance plant development, warrant further thought. The biodegradation of persistent organic pollutants (POPs) can be improved through the course of co-digestion by giving cometabolites or inducers. This approach has been exhibited in various examinations, which have shown that the utilization of engineered or regular inducers, for example, biphenyl, terpenes, linoleic corrosive, and salicylate, can fundamentally build the biodegradation of different POPs.
3. Moreover, the presentation of endophytic microbes, like *Rhizobium*, into defiled soil can possibly advance plant development and improve the presence of debasing microorganisms and saprotrophic growths. For instance, double immunization of horse feed with *G. caledonium* and *Rhizobium meliloti* has been displayed to have incredible potential for the remediation of PCB-debased soil. Besides, joined treatment utilizing a blend of arbuscular mycorrhizal fungi (AMF), *Sphingomonas paucimobilis* microorganisms, and the saprotrophic growth *Cunninghamella echinulata* have been found to debase more oil than a solitary vaccination of ryegrass with AMF alone or microbes or saprotrophic growths. A new report (Lu & Lu, 2015) detailed that the scattering of polycyclic sweet-smelling hydrocarbons (PAHs) and polychlorinated biphenyls (PCBs) was more prominent in the consolidated treatment of plants, AMF, and worms. A joined technique that uses biosurfactants with co-immunization of microbes and AMF might be successful for the remediation of old soil sullied with POPs (Zhou et al., 2013).

8.5.5 *Plant-Based BR (Bioremediation)*

The concept of "phytoremediation" was first introduced in the 1990s as a method for utilizing plants to remove toxic metals (Raskin et al., 1994; Salt et al., 1995). Since then, this technology has been widely applied to effectively remove a wide range of organic and inorganic pollutants (Pilon-Smits, 2005). One of the key advantages of phytoremediation is its reliance on solar energy, making it a cost-effective option. Additionally, it does not necessitate the removal of soil, allowing for the preservation of the unique soil after treatment. Besides, phytoremediation is appropriate for in situ debasement of poisons over a huge region, further expanding its expense viability. The bioavailability of toxins is a vital figure for their remediation, which is subject to the synthetic properties of the contamination, soil properties, natural circumstances, and organic action. One significant synthetic property that influences the development of poisons in soil is hydrophobicity. Steady natural poisons (POPs) are exceptionally hydrophobic atoms that firmly will undoubtedly soil natural matter and don't disintegrate in soil pore water (Briggs et al., 1982). This absence of bioavailability restricts their ability for phytoremediation and characterizes them as diligent contaminations. Plants that can retain high groupings of poisons, known as hyperaccumulators, are fundamental for the compelling expulsion of contaminations in soil. While some plant species have been explored for the remediation of normal natural poisons, there have been reports of take-up or resilience to POPs like DDT, PCBs, and HCH in different plant species.

8.5.5.1 A Transgenic Approach to POP Degradation

The phytoremediation of persistent organic pollutants (POPs) poses a significant challenge due to the difficulty in decomposing these compounds into nontoxic forms. This is further complicated by the fact that some POPs can also be toxic to plant cells. Therefore, it is essential that POPs are rapidly degraded after uptake in order for phytoremediation to be effective. One strategy to achieve this is through the use of endogenous enzymatic activities within plant cells. Another approach is the utilization of catabolic compounds separated from bacterial strains that have the capacity to corrupt POPs. There are a few instances of competitor qualities, for example, those encoding biphenyl dioxygenase (BphA qualities), which have been secluded from *Rhodococcus* sp. also, have been displayed to debase polychlorinated biphenyls (PCBs) (Masai et al., 1995).

Moreover, bioaugmentation, which includes the immediate utilization of these microorganisms to debased soil, has been proposed as an answer (Singer & Fehr, 2005). Nonetheless, the reasonability of microbes presented in this way can be impacted by a scope of elements, like temperature, dampness, pH, and natural matter substance. An elective technique is the utilization of transgenic plants that express at least one protein fit for corrupting POPs, which presents an easy route to delivering plants for phytoremediation of POPs. Studies have announced the retention and detoxification of natural poisons, for example, explosives and herbicides utilizing transgenic plants (Vail et al., 2014). Besides, the biodegradation of atrazine has been accounted for in transgenic kinds of grass and horse feed that express adjusted bacterial qualities for atrazine chlorohydrolase.

8.6 Conclusion

The primary endorsed approaches for the remediation of persistent organic pollutants (POPs) include both phytoremediation and microbial remediation. Phytoremediation involves the use of plants for the removal of pollutants from the environment and is powered by solar energy, making it a cost-effective option. However, the discovery of more hyperaccumulator plant species is needed to meet the increasing demand for phytoremediation. Microbial remediation, on the other hand, involves the use of microorganisms to degrade pollutants and is often implemented through the use of biofilms, which are communities of microorganisms that form on surfaces. Biofilm-based remediation innovation is one of the more successful, eco-accommodating, and simple-to-use eliminating poisons from the climate. Microbial biofilms are progressively being utilized as a pointer framework for checking the extremely weighty metal pollution which is in water. Moreover, the utilization of transgenic plants and organisms plant blends has additionally been concentrated as potential remediation approaches for POPs. Transgenic plants have been viewed as a promising methodology for the bioremediation of dynamically tainted soils, yet further examination is expected to comprehend their potential completely. Parasite plant blends have been viewed as a promising methodology for the bioremediation of dynamically debased soils. In any case, the utilization of transgenic plants and organism plant blends has not been generally embraced as remediation methods yet.

For choosing the most suitable remediation plan, it is especially urgent to think about the top most substance properties of the poison, soil properties, ecological circumstances, and organic action. One significant compound property of an or in the poison that influences its development in soil is its actual hydrophobicity. Constant natural poisons (POPs) are exceptionally hydrophobic atoms that are firmly bound to the dirt natural matter and don't disintegrate in soil pore water. This absence of bioavailability restricts their ability for phytoremediation, making them delegated tireless contaminations.

This chapter provides a comprehensive overview of the primary endorsed approaches for POP remediation, as well as the dominant components utilized in the degradation of natural contaminants in water and soil. Furthermore, it presents an evaluation of the advantages of various remediation techniques, examines POPs, and explores the method of selecting the most appropriate remediation plan to degrade various types of POPs, filling the knowledge gap in this area, and providing guidance for future research and applications of POPs.

8.7 Future Prospects

In addition to phytoremediation and microbial remediation, other approaches such as bioremediation, bioventing, and biostimulation have also been researched and implemented for the remediation of persistent organic pollutants (POPs). Bioremediation involves the use of microorganisms to degrade pollutants, often through the use of genetically modified microorganisms or the addition of nutrients to stimulate the growth of indigenous microorganisms. Bioventing and biostimulation involve the use of air or nutrients, respectively, to enhance the activity of microorganisms in the soil.

Another approach that has been gaining attention in recent years is the use of nanotechnology for the remediation of POPs. This involves the use of nanoparticles to adsorb or degrade pollutants and has been found to be effective in laboratory studies. However, more research in today's world is needed to fully understand the potential and limitations of this approach, as well as its potential risks to human health and the environment.

In order to effectively remediate POPs, a combination of different approaches may be necessary, as well as an understanding of the specific contaminants present and the characteristics of the contaminated site. Furthermore, the use of phytoremediation, microbial remediation, and other techniques should be carefully evaluated and monitored to ensure that the remediation process is safe and effective.

In conclusion, while microbial remediation has been extensively researched, comparatively little research has been conducted on the use of phytoremediation for the remediation of POPs. To meet the growing demand for phytoremediation, it is crucial to discover more hyperaccumulator plant species. Furthermore, a combination of different approaches such as bioremediation, bioventing, biostimulation, nanotechnology, and other techniques may be necessary to effectively remediate POPs.

References

Alharbi, O. M., Khattab, R. A., & Ali, I. (2018). Health and environmental effects of persistent organic pollutants. *Journal of Molecular Liquids, 263*, 442–453.

Ashraf, M. A., Yusoff, I., Yusof, M., & Alias, Y. (2013). Study of contaminant transport at an open-tipping waste disposal site. *Environmental Science and Pollution Research, 20*(7), 4689–4710.

Authman, M. M., Zaki, M. S., Khallaf, E. A., & Abbas, H. H. (2015). Use of fish as bio-indicator of the effects of heavy metals pollution. *Journal of Aquaculture Research & Development, 6*(4), 1–13.

Berg, D. A., Skillman, E. D., Henry, R. B., Erb, D. K., & Carigi, L. (2016). Carbon and oxygen abundances in low metallicity dwarf galaxies. *The Astrophysical Journal, 827*(2), 126.

Briggs, G. G., Bromilow, R. H., & Evans, A. A. (1982). Relationships between lipophilicity and root uptake and translocation of non-ionised chemicals by barley. *Pesticide Science, 13*(5), 495–504.

Čvančarová, M., Moeder, M., Filipová, A., & Cajthaml, T. (2015). Biotransformation of fluoroquinolone antibiotics by ligninolytic fungi–metabolites, enzymes and residual antibacterial activity. *Chemosphere, 136*, 311–320.

Damstra, T. (2002). Potential effects of certain persistent organic pollutants and endocrine disrupting chemicals on the health of children. *Journal of Toxicology: Clinical Toxicology, 40*(4), 457–465.

Dewan, P., Jain, V., Gupta, P., & Banerjee, B. D. (2013). Organochlorine pesticide residues in maternal blood, cord blood, placenta, and breastmilk and their relation to birth size. *Chemosphere, 90*(5), 1704–1710.

Enerijiofi, K. E. (2021). Bioremediation of environmental contaminants: A sustainable alternative to environmental management. In *Bioremediation for environmental sustainability* (pp. 461–480). Elsevier.

El-Shahawi, M. S., Hamza, A., Bashammakh, A. S., & Al-Saggaf, W. T. (2010). An overview on the accumulation, distribution, transformations, toxicity and analytical methods for the monitoring of persistent organic pollutants. *Talanta, 80*(5), 1587–1597.

Friedrich, T., Timmermann, A., Tigchelaar, M., Elison Timm, O., & Ganopolski, A. (2016). Nonlinear climate sensitivity and its implications for future greenhouse warming. *Science Advances, 2*(11), e1501923.

Hayashi, O., Kameshiro, M., & Satoh, K. (2013). Degradation of the persistent organic pollutant heptachlor in Japanese field soils. *Bioscience, Biotechnology, and Biochemistry., 77*(6), 1240–1244.

Holoubek, I. (2011). *Environmental fate of persistent organic pollutants definition, sources, fate, recetox.*

Hussain, A., Rehman, F., Rafeeq, H., Waqas, M., Asghar, A., Afsheen, N., et al. (2022). In-situ, Ex-situ, and nano-remediation strategies to treat polluted soil, water, and air–A review. *Chemosphere, 289*(133), 252.

Kearney, C. P., & Kaufaman, D. D. (1973). *Microbial degradation of some chlorinated pesticides.*

Khezri, B., Fisher, A. C., & Pumera, M. (2017). CO_2 reduction: The quest for electrocatalytic materials. *Journal of Materials Chemistry A, 5*(18), 8230–8246.

Lu, Y. F., & Lu, M. (2015). Remediation of PAH-contaminated soil by the combination of tall fescue, arbuscular mycorrhizal fungus and epigeic earthworms. *Journal of hazardous materials, 285*, 535–541.

Masai, E., Yamada, A., Healy, J. M., Hatta, T., Kimbara, K., Fukuda, M., & Yano, K. (1995). Characterization of biphenyl catabolic genes of gram-positive polychlorinated biphenyl degrader Rhodococcus sp. strain RHA1. *Applied and Environmental Microbiology, 61*(6), 2079–2085.

Pilon-Smits, E. (2005). Phytoremediation. *Annual Review of Plant Biology, 56*, 15–39. https://doi.org/10.1146/annurev.arplant.56.032604.144214. PMID: 15862088.

Raskin, I., Kumar, P. N., Dushenkov, S., & Salt, D. E. (1994). Bioconcentration of heavy metals by plants. *Current Opinion in Biotechnology, 5*(3), 285–290.

Ritter, L., Solomon, K. R., Forget, J., Stemeroff, M., & O'Leary, C. (2007). *Persistent organic pollutants*. United Nations Environment Programme.

Salt, D. E., Blaylock, M., Kumar, N. P., Dushenkov, V., Ensley, B. D., Chet, I., & Raskin, I. (1995). Phytoremediation: a novel strategy for the removal of toxic metals from the environment using plants. *Biotechnology, 13*(5), 468–474.

Sieverding, E., da Silva, G. A., Berndt, R., & Oehl, F. (2014). Rhizoglomus, a new genus of the Glomeraceae. *Mycotaxon, 129*(2), 373–386.

Singer, T., & Fehr, E. (2005). The neuroeconomics of mind reading and empathy. *American Economic Review, 95*(2), 340–345.

Tegene, B. G., & Tenkegna, T. A. (2020). Mode of action, mechanism and role of microbes in bioremediation service for environmental pollution management. *Journal of Biotechnology & Bioinformatics Research*. SRC/JBBR-112, *116*(2), 39–50. https://doi.org/10.47363/JBBR/2020

Urbaniak, M. (2013). Biodegradation of PCDDs/PCDFs and PCBs.

Vail, A. L., Manica, A., & Bshary, R. (2014). Fish choose appropriately when and with whom to collaborate. *Current Biology, 24*(17), R791–R793.

Wang, S., Zhang, S., Huang, H., & Christie, P. (2011). Behavior of decabromodiphenyl ether (BDE-209) in soil: effects of rhizosphere and mycorrhizal colonization of ryegrass roots. *Environmental Pollution, 159*(3), 749–753.

Wang, F. M., Lesnikowski, A., Ford, J. D., & Biesbroek, G. R. (2022). UNEP 2017. The Adaptation Gap Report 2017: Towards global assessment. United Nations Environment Programme (UNEP).

Ying, J., Li, H., Seng, T. J., Langford, C., Srivastava, G., Tsao, S. W., ... & Tao, Q. (2006). Functional epigenetics identifies a protocadherin PCDH10 as a candidate tumor suppressor for nasopharyngeal, esophageal and multiple other carcinom as with frequent methylation. *Oncogene, 25*(7), 1070–1080.

Zacharia, J. T., & Croatia, R.. (2011). *Pesticides in the modern world—Risks and benefits*, vol. 2, Chapter 7 (Book) Intech; ISBN 978-953-307-437-5

Zhou, X., Zhou, J., Xiang, X., Cébron, A., Béguiristain, T., & Leyval, C. (2013). Impact of four plant species and Arbuscular Mycorrhizal (AM) Fungi on Polycyclic Aromatic Hydrocarbon (PAH) dissipation in spiked soil. *Polish Journal of Environmental Studies, 22*(4).

Chapter 9
An Overview of Heavy Metal Pollution in Aquatic Sediments Around the World

Princy M. John and Anu Gopinath

9.1 Introduction

Numerous studies have confirmed that marine sediments are highly polluted by heavy metals; as a result, assessing the distribution of metals in surface sediment is important for evaluating contamination in the marine environment. Sediments are the key repository and source of metals in the aquatic environment, and they play a crucial role in the transportation and storage of potentially harmful metals. The industrial waste reaching the sea via atmospheric precipitation and dumping of urban and rural waste is mostly responsible for the input of heavy metals into the marine environment which are successively assimilated into the sediments.

9.2 Heavy Metal Pollution Reported from Aquatic Sediments During the Period 2000–2010

An investigation conducted in sediments of central-southwest coast of India revealed that the period immediately following the summer monsoon was shown to have the highest metal accumulation due to the flushing of previously deposited sediments from the Cochin backwaters and the weak coastal currents encouraging deposition in the coastal environment. During the other two seasons, cross-shelf transport was

P. M. John
Faculty of Ocean Science and Technology, Kerala University of Fisheries and Ocean Studies, Cochin, India

A. Gopinath (✉)
Department of Aquatic Environment Management, Kerala University of Fisheries and Ocean Studies, Cochin, India

H. Alshemmari et al. (eds.), *Contaminated Land and Water*, Emerging Contaminants and Associated Treatment Technologies, https://doi.org/10.1007/978-3-031-65129-8_9

determined to be minimal. The geochemical studies reveal two processes that control the regional distribution of trace elements in this environment: (1) increased monsoonal supply of materials through the Cochin backwaters, resulting in increased metal deposition; and (2) effective masking of these metals through incorporation into clay and organic association, which occurs after the monsoon (Balachandran et al., 2003).

Laluraj and Nair (2006) concluded that the concentrations of fairly redox-insensitive trace metals including Cu, Ni, Zn, Cr, and Cd (except Cr) were greater at nearshore sediments of the west coast of the Arabian Sea, then fell as they moved further out to sea, and finally increased slightly at oxygen minimum sites. Because the scavenging or releasing effects of Fe and Mn oxides/hydroxides act as significant "sinks" or "sources" of heavy metals, this geochemical diversity in their distributional properties is essentially connected with the amount to which Fe and Mn oxides/hydroxides precipitate or dissolve. The reducing condition may arise as a result of high productivity and organic matter in the sediment, both of which are particularly effective in removing and adsorbing a variety of elements from the sediments.

As per the study carried out in the coastal sediments of the Arabian Sea along with the urban Karachi (Pakistan) region, sediment-bound heavy metal concentrations are highest at the Arabian Sea's convergence of the Lyari and Malir river streams, showing the impact of Karachi's heavily urbanized and industrialized effluents. Furthermore, sediment quality testing found that the marine environment's potential toxicity could have negative biological consequences for the biota directly and indirectly through the food chain. It is also stated that human activities and urbanization are the primary causes of heavy metal contamination in the Karachi region's maritime environment (Siddique et al., 2009).

Similarly, an investigation carried out in the Gulf of Mannar, southeast coast of India by Jonathan et al. (2004), revealed that metals are mostly emitted by industrial waste effluents on the northern side of the Gulf of Mannar and river discharge on the southern side. The area, while contaminated, is not in great danger because most toxic metals are below the levels that cause toxic effects on marine organisms exposed to the sediments, and studies conducted at many contaminated sites have shown that sediment quality can be restored with proper environmental strategies.

In the last three decades, the rapid economic expansion in the Pearl River Delta (PRD) region of South China has had a considerable impact on the local ecosystem. Estuarine sediments serve as a key sink for pollutants and nutrients in the environment. Trace metal deposition in sediments has the potential to produce major environmental concerns in the aquatic system. When compared to the sub-surface sediment layers, the concentrations of Cu, Pb, and Zn in the surface sediment layers were generally higher. Since the 1970s, the PRE has seen a large rise in Cu, Pb, and Zn imports. Additionally, different sampling locations in the estuary got somewhat varied sorts of inputs, according to the findings (Ip et al., 2004).

As per Ip et al. (2007), Cu, Cr, Pb, and Zn distributions in Pearl River Estuary showed a typical diffusion trend from the land to the sea. The mixed zone between freshwater and marine waters had two hotspots of trace metal contamination. The deposition of dissolved and particulate trace metals in the water column at the estuarine area could be responsible for the metal enrichment in the sediments. The Pearl River Estuary and its surrounding coastal area have similar Pb isotopic signatures, indicating that the Pearl River Estuary was a key supply of trace metals to the area.

Cd, Zn, and Ni were found in high concentrations in the sediments of Pearl River Estuary coastal wetlands, according to Q. S. Li et al. (2007). The sediment quality in mangrove swamps and mudflats was unsuitable for a nature reserve, aquaculture, or reclamation project. In addition, the chemical partitioning of Cd and Zn was primarily related to exchangeable fractions, which raised the ecological danger. Because of the low environmental quality in the area, the current coastal wetland utilization practices in this region should be revisited.

Preda and Cox (2002) concluded that the geological bedrock provides all of the metals present in the sediments of the Pumicestone region, Southeast Queensland, Australia. In sediments, Zn, V, and Cr are the most common trace elements found. The remaining metals have a lot of variation in terms of their spatial distribution. The presence of Fe and Mn oxides, as well as the sediment grain size, determines the concentration of all trace elements. Fine-grained Fe-rich materials absorb more trace metals than sandy soils in general. Some metals, such as Cr, Mo, Pb, and As, are found in higher concentrations in soils that have evolved from estuary muds.

Also, Cox and Preda (2005) investigated the content and distribution of trace metals in sediments in the subtropical setting of Moreton Bay on Australia's east coast. When compared to the findings of similar studies in other densely populated areas of Australia, the Moreton Bay area sediments have a relatively low metal content. Concentrations in fine-grained estuary sediments are typically two to three times greater than in sandy material along the Bay shoreline. The elements most typically raised and warranting additional investigation when compared to Australian and New Zealand Environment and Conservation Council Guidelines for metal concentrations in sediments include Ni, Hg, Pb, Zn, and on rare occasions Cd, As, and Cr. Estuaries near highly populous or industrialized areas, such as those around the Brisbane River, have the highest metal concentrations.

The natural and anthropogenic mechanisms that govern sediment chemistry in Admiralty Bay, King George Island, Antarctica, are investigated by Santos et al. (2005), with a focus on the area affected by the Brazilian Antarctic Station Comandante Ferraz (Ferraz). Because of the small area of polluted sediments and the low bioavailability of heavy metals, this is unlikely to harm Antarctic benthic marine species. Moreover, contamination levels are lower than those found in other Antarctic Research Stations and developed areas. For identifying and preventing pollution in Admiralty Bay, continuous environmental monitoring, assessment of local background through sediment cores, transference rates to biota, and additional ecotoxicological investigations are required.

Most elements in the studied coastal sediments from Admiralty Bay, maritime Antarctica, had high redox potential gradients and limited vertical variability. Rare earth and other element concentrations are comparable to the dominating local rocks. These findings suggest that postdepositional activities have little impact on the geochemistry of the elements studied and that local bedrock is the primary source of sediments (Santos et al., 2007).

According to Massimo Gasparon and Matschullat (2006), several types of human impact have been identified and described in the Larsemann Hills, East Antarctica, including eutrophication of certain lake systems; elevated concentrations of N species and hydrocarbons in the immediate vicinity of some station buildings; metal, plastic, paper, and timber debris carried by the wind; and construction activities required for the station's establishment. The only evident human impact on the Larsemann Hills regolith is an enhanced rate of erosion and weathering in the proximity of heavy traffic regions—vehicular tracks and station areas—when these small significantly damaged sites are excluded.

9.3 Heavy Metal Pollution Reported from Aquatic Sediments During the Period 2010–2021

The heavy metal pollution reported from aquatic sediments during the period 2010–2021 is shown in Table 9.1. An investigation carried out in Bohai Bay, China, by Gao and Chen (2012) revealed that heavy metal enrichment was found in abundance in the coastal sediments of the Bohai Bay. The sediment pollution in this area was mostly caused by nonpoint source inputs from both river transport and atmospheric deposition. Land-based point source discharges could be to blame for the very high amounts in particular riverine areas. Except for Cd at all locations and Zn at some riverine stations, the metals analyzed exhibit minimal mobility. Despite the fact that Cd has high mobility, due to its low overall concentrations, it is unlikely to have a negative impact on the environment. Although there was a clear anthropogenic influence on the accumulation of examined heavy metals in Bohai Bay sediments, their levels were still relatively modest compared to some other marine coastal areas that receive significant anthropogenic inputs. Similarly, Gao and Li (2012) assessed six trace metals (Cd, Cr, Cu, Ni, Pb, and Zn) geochemically and environmentally in surface sediments from intertidal Bohai Bay. The findings show that sediment grain size has a significant impact on their distribution and fractionation. Clayey silt sediments have significantly greater metal contents than sand and silty sand sediments. Cd and Pb are more mobile in clayey silt sediments than in sand and silty sand sediments.

The Bohai Sea is a semi-closed sea with limited water exchange capacity that has been regarded as one of China's most contaminated areas and has received public attention in recent decades. Rapid industrialization and urbanization in the coastal region have resulted in serious pollution pressure in the Bohai Sea in recent years. Hg and Cd contents were much higher than the typical upper crust value, posing a high potential ecological danger and a significant potential ecological risk, respectively. According to the marine sediment quality of China, the sedimentary

Table 9.1 Metal concentrations (mg/kg dry wt) in the surface sediments of different marine environments collected in the recent 21 years. Values in the table are the reported range of concentrations

Region	Cu	Ni	Pb	Zn	Fe	Co	Mn	Cr	References
Pumicestone region, Southeast Queensland, Australia	ND-8	ND-14	ND-33	ND-206	–	ND-16	–	1–91	Preda and Cox (2002)
Central-southwest coast of India	–	0–213	0–71.3	–	–	0–40	–	0–205	Balachandran et al. (2003)
Gulf of Mannar, southeast coast of India	34–110	14–58	9–21	39–162	–	9–29	–	71–963	Jonathan et al. (2004)
Moreton Bay, Southeast Queensland, Australia	0–31	0–20	2–59	5–89	1264–113,925	–	3–322	1–331	Preda and Cox (2002)
West coast of Arabian Sea	–	–	0.50–3.04	–	–	4.41–31.93	–	–	Laluraj and Nair (2006)
Larsemann Hills, East Antarctica	0.8–6.5	0.1–159	7.1–62.2	7.8–200	–	0.6–62	–	0.1–266	Massimo Gasparon and Matschullat (2006)
Pearl River estuary, China	6.2–100	10.6–54.1	16.0–96.3	55.1–268	–	7.4–24.0	–	33.8–135	Ip et al. (2007)
Pearl River estuary, China	ND-351.2	13.8–318.3	12.3–86.0	120.5–478.4	–	–	–	13.5–191.2	Li et al. (2007)
Admiralty Bay, King George Island, Antarctica	–	–	–	24.64–142.53	23–978	0.64–22.86	–	2.07–52.77	Santos et al. (2007)
Karachi coastal region, Pakistan	6.95–272	7.48–74.91	6.33–121.03	3.31–389.23	7100–64,700	5.14–95.01	1.15–318	2.95–571	Siddique et al. (2009)

(continued)

Table 9.1 (continued)

Region	Cu	Ni	Pb	Zn	Fe	Co	Mn	Cr	References
Coastal Bohai Bay, China	20.1–62.9	23.4–52.7	20.9–66.4	55.3–457.3	–	–	–	60.1–224.5	Gao and Chen (2012)
Intertidal Bohai Bay, China	7.9–46.7	14.1–47.9	18.8–39.1	34.0–123	–	–	–	36.7–110	Gao and Li (2012)
Kongsfjorden,Svalbard	17.7–22.8	22.9–31.3	14.3–19.9	66.6–78.6	–	10.4–14.6	361–532	57.3–72.2	Grotti et al. (2013)
Deception Bay, Queensland, Australia	1.9–20.8	6.5–36.7	ND-13.3	5.5–548	3829–46,160	ND- 13	54.5–685	107–233	Brady et al. (2014)
Bramble Bay, Queensland, Australia	3.54–39.29	8.96–39.21	2.218–22.46	12.38–121.8	8940–48,180	4.455–25.42	126.3–826	26.58–159.5	Brady et al. (2015)
Central west coast of India	–	–	0.1–163.3	–	–	–	–	–	Chakraborty et al. (2015)
Al-Kharrar lagoon, Red Sea, Saudi Arabia	11–28	7.9–9.8	40–63	30–51	1708–3686	2.2–8.1	53–225	14–25	Youssef and El-Sorogy (2016)
Al-Kharrar lagoon, Red Sea, Saudi Arabia	4.2–43.06	0.0–56.76	0.0–1.27	0.19–46.28	1585.7–32,501.9	0.0–15.41	77.7–510.1	–	Hariri and Abu-Zied (2018)
Pearl River estuary, China	5.8–170.6	–	23–78	32–259	–	–	–	12–130	Zhao et al. (2017)
Bohai Bay, China	23–31	–	18.8–28.4	56.7–83.4	–	–	–	39–72.6	Li et al. (2018)
Bohai Sea, China	2.22–38.94	2.89–48.25	8.93–37.67	6.06–129.60	–	–	–	5.61–110.60	Zhu et al. (2020)
Southwest coast of India	11.2– 18.2	32.6–67.6	8.1–31	16.8–39	8573.8–22,118	4.8–13.5	74.7–219	107–165	Cyriac et al. (2021)

ecosystems of coastal Bohai were in good condition, with the exception of Jinzhou Bay (Li et al., 2018).

In addition, Zhu et al. (2020) found that the spatial distributions of Cu, Pb, Ni, Cr, and Zn were found to be similar in Bohai Bay. The concentrations of these metals were highest in this region, which is characterized by fine-grained sediments. Cd, Hg, and As had unique geographical distributions from each other and from other heavy metals studied. The ecological risk linked with measured Cd and Hg levels was significant in Jinzhou Bay and Fuzhou Bay, indicating a high potential for ecosystem degradation.

The research by Zhao et al. (2017) in Pearl River Estuary, China, found that silt and sandy silt dominated the surface sediments in the study area, indicating strong hydrodynamic conditions. Heavy metal concentrations had a close relationship with sediment type in terms of spatial distribution. Overall, these sediments met China's primary standard criteria for Cd, Cr, Hg, Pb, and Zn, but not for Cu (CSBTS, 2002). Within the compositional range of other Chinese intertidal sands, heavy metal concentrations increased.

The research carried out in Admiralty Bay, Antarctica, by Trevizani et al. (2016) showed that high quantities of Cu and Zn were observed throughout Admiralty Bay which can be linked to the region's geological composition. The current study's findings are critical for monitoring and studying heavy metal contamination in Antarctica, as well as providing a research foundation for the Admiralty Bay environment.

According to Sun et al. (2013), anthropogenic influences on Zn and Cd in the sediments of Prydz Bay are minimal. The principal sources of trace elements are biogenic and lithogenic inputs, which are about equal for Zn and Cd at the AIS margin. The continental shelf and deep ocean are dominated by biogenic input, which is substantially higher for Zn than for Cd in the deep ocean. Three types of relationships between biogenic Zn and biogenic Cd have been identified, representing phytoplankton's biological absorption of trace elements.

Surface sediments from Prydz Bay, East Antarctica, were studied to learn more about their properties, depositional processes, source, and productivity in the past. The grain size distribution suggests that sediment transport was away from the coast, with high energy conditions near the beach indicating a large detrital input. Lithogenic nature was discovered in major metals such as Al, Ti, Fe, Mn, and Mg. Trace metals were derived from both biogenic and lithogenic sources, but the biogenic input was more prevalent, indicating considerable biological activity in the area (Choudhary et al., 2018).

A previous study by Monien et al. (2014) found that the HNLC zones of the Southern Ocean depend greatly on iron from the suboxic shelf sediments. Increased glacier retreat and melt water fluxes, as well as the resulting rise in accumulation of iron-rich sediment on the shelf, suggest that the role of Antarctic shelf sediments in natural iron fertilization of the Southern Ocean may become more important in the future. This may be especially true for sediments from shallow shelf areas, because increasing melting may also enhance stratification, which might cut off Fe supplies from deeper seas for broad parts of the shelf, making shallow sediment sources even more significant.

As per Grotti et al. (2013), total concentration values and solid speciation patterns of trace elements in the sediment samples from Kongsfjorden, Svalbard, suggested that the anthropogenic impact of trace elements in the studied area is insignificant, though a few elements showed a modest enrichment compared to crustal values (As, Cd, Cr, Ni, and V). Furthermore, it was obvious that the bioavailable fraction of each element that may enter the aquatic environment makes up a modest portion of the total concentration. This emphasizes the danger of overestimating the environmental impact of heavy metals when only total concentrations are considered.

Chakraborty et al. (2015) observed that Pb contamination was found along India's northeast, northwest, and southwest coasts. The Fe-Mn oxyhydroxide mineral phases of the sediments regulated the distribution of Pb in sediments from India's northeast and northwest coasts. Organic carbon (OC) seems to be the most important element in influencing Pb distribution in sediments from India's central east and southwest coasts. The findings of this research could aid decision-making in terms of predicting Pb levels from natural and anthropogenic sources, as well as controlling Pb pollution in coastal and estuarine sediments throughout India.

The findings by Cyriac et al. (2021) revealed that particle size, organic carbon, and Fe are the key determinants of trace metal distribution in surface sediments along the southwest coast of India. The studied region appears to be moderately contaminated by Ni, Cr, and Pb, according to the geochemical indices (contamination factor, enrichment factor, and geo-accumulation factor). Pb, Cr, and Ni enrichment may have adverse implications on benthic creatures, according to ecological risk assessment standards. In addition to the foregoing, an increase in organic matter enrichment causes higher trace metal deposition in sediments.

The research by Youssef and El-Sorogy (2016) looked at the spatial distribution and levels of heavy metal contamination in the bottom sediments of the Al-Kharrar coastal lagoon near the Red Sea in Saudi Arabia. Heavy metal concentrations varied regionally, although distribution patterns were rather consistent. Contamination is mostly caused by air intake and diffusion from anthropogenic activities such as dredging and reclamation. The pollution stress in the research area is not very high, yet the sediments are categorized as Pb-polluted. According to the Igeo index, the sediments of Al-Kharrar lagoon are moderately polluted with Pb, Cr, Cu, and As levels. The high EF values for Pb, As, Cu, Cr, Co, and Zn imply that metal concentrations in bottom sediments are anthropogenic. The CF results show that the Al-Kharrar lagoon's bottom sediments are significantly contaminated with Mn and Pb.

Also, an investigation carried out in Al-Kharrar Lagoon by Hariri and Abu-Zied (2018) showed that metals were directly correlated with Fe and Mn in the Al-Kharrar Lagoon, but there were no relationships with organic carbon due to the abundance of Fe-Mn oxides and highly oxygenated waters, which could affect the role of organic matter in forming complexes easily for fixation and precipitation of heavy metals into the lagoon's sediments. All metals had favorable associations with dissolved oxygen and muddy sediments in this study.

As per the study conducted in 13 sediment samples of Deception Bay, Queensland, Australia, it represents that lead has contaminated most of the locations, but there has also been an enrichment of arsenic, zinc, and other heavy metals. The modified degree of contamination indices indicates there is little contamination; however, the Nemerow pollution index indicates Deception Bay is substantially contaminated (Brady et al., 2014).

According to Brady et al. (2015), the general sediment health of Bramble Bay was determined to be good, except for Ni and Cr, which exceeded the Australian Interim Sediment Quality Guidelines. This finding, as well as the presence of anthropogenic Hg, necessitates more inquiry to ensure Bramble Bay's health is preserved.

9.4 Conclusions

Heavy metals are found in natural levels in aquatic sediments. As a result, their sheer presence cannot be interpreted as a sign of contamination. Furthermore, even high heavy metal concentrations are not always harmful to the environment or human health. This review highlights the literature on heavy metals in surface sediments of aquatic environments from the 2000s to the 2021s. The impact of coastal human activities on heavy metal concentrations and dispersion in global ocean sediments was growing. Heavy metal contents in aquatic surface sediments have increased in the recent decades. However, the concentration of heavy metals in the Bohai Sea and Pearl River Estuary from China has decreased considerably. Heavy metal distributions were closely associated with surface sediment properties. Riverine input, atmospheric deposition, and emissions from coastal human activities together contributed to the presence of heavy metals in the world's oceans. The industrial waste reaching the sea via atmospheric precipitation and dumping of urban and rural waste is mostly responsible for the input of heavy metals into the marine environment which is successively assimilated into the sediments. Environmental monitoring programs should include sediments since they are key hosts for pollution heavy metals.

References

Balachandran, K. K., Joseph, T., Nair, M., Sankaranarayanan, V. N., Kesavadas, V., & Sheeba, P. (2003). Geochemistry of surficial sediments along the central southwest coast of India—Seasonal changes in regional distribution. *Journal of Coastal Research, 19*(3), 664–683.

Brady, J. P., Ayoko, G. A., Martens, W. N., & Goonetilleke, A. (2014). Enrichment, distribution and sources of heavy metals in the sediments of Deception Bay, Queensland, Australia. *Marine Pollution Bulletin, 81*(1), 248–255. https://doi.org/10.1016/j.marpolbul.2014.01.031

Brady, J. P., Ayoko, G. A., Martens, W. N., & Goonetilleke, A. (2015). Weak acid extractable metals in Bramble Bay, Queensland, Australia: Temporal behaviour, enrichment and source apportionment. *Marine Pollution Bulletin, 91*(1), 380–388. https://doi.org/10.1016/j.marpolbul.2014.11.048

Chakraborty, S., Chakraborty, P., & Nath, B. N. (2015). Lead distribution in coastal and estuarine sediments around India. *Marine Pollution Bulletin, 97*(1–2), 36–46. https://doi.org/10.1016/j.marpolbul.2015.05.056

Choudhary, S., Nayak, G. N., Tiwari, A. K., & Khare, N. (2018). Source, processes and productivity from distribution of surface sediments, Prydz Bay, East Antarctica. *Polar Science, 18*(June), 63–71. https://doi.org/10.1016/j.polar.2018.06.002

Cox, M. E., & Preda, M. (2005). Trace metal distribution within marine and estuarine sediments of western Moreton Bay, Queensland, Australia: Relation to land use and setting. *Geographical Research, 43*(2), 173–193. https://doi.org/10.1111/j.1745-5871.2005.00312.x

CSBTS (China State Bureau of Quality and Technical Supervision) (2002). The People's Republic of China National Standards GB 18668-2002 –Marine Sediment Quality. p. 10.

Cyriac, M., Gireeshkumar, T. R., Furtado, C. M., Fathin, K. P. F., Shameem, K., Shaik, A., Vignesh, E. R., Nair, M., Kocherla, M., & Balachandran, K. K. (2021). Distribution, contamination status and bioavailability of trace metals in surface sediments along the southwest coast of India. *Marine Pollution Bulletin, 164*(January), 112042. https://doi.org/10.1016/j.marpolbul.2021.112042

Gao, X., & Chen, C. T. A. (2012). Heavy metal pollution status in surface sediments of the coastal Bohai Bay. *Water Research, 46*(6), 1901–1911. https://doi.org/10.1016/j.watres.2012.01.007

Gao, X., & Li, P. (2012). Concentration and fractionation of trace metals in surface sediments of intertidal Bohai Bay, China. *Marine Pollution Bulletin, 64*(8), 1529–1536.

Gasparon, M., & Matschullat, J. (2006). Geogenic sources and sinks of trace metals in the Larsemann Hills, East Antarctica: Natural processes and human impact. *Applied Geochemistry, 21*(2), 318–334. https://doi.org/10.1016/j.apgeochem.2005.09.013

Grotti, M., Soggia, F., Ianni, C., Magi, E., & Udisti, R. (2013). Bioavailability of trace elements in surface sediments from Kongsfjorden, Svalbard. *Marine Pollution Bulletin, 77*(1–2), 367–374. https://doi.org/10.1016/j.marpolbul.2013.10.010

Hariri, M. S. B., & Abu-Zied, R. H. (2018). Factors influencing heavy metal concentrations in the bottom sediments of the Al-Kharrar lagoon and Salman Bay, eastern Red Sea coast, Saudi Arabia. *Arabian Journal of Geosciences, 11*(17). https://doi.org/10.1007/s12517-018-3838-2

Ip, C. C., Li, X. D., Zhang, G., Wai, O. W., & Li, Y. S. (2007). Trace metal distribution in sediments of the Pearl River Estuary and the surrounding coastal area, South China. *Environmental Pollution, 147*(2), 311–323.

Ip, C. C. M., Li, X. D., Zhang, G., Farmer, J. G., Wai, O. W. H., & Li, Y. S. (2004). Over one hundred years of trace metal fluxes in the sediments of the Pearl River Estuary, South China. *Environmental Pollution, 132*(1), 157–172. https://doi.org/10.1016/j.envpol.2004.03.028

Jonathan, M. P., Ram-Mohan, V., & Srinivasalu, S. (2004). Geochemical variations of major and trace elements in recent sediments, off the Gulf of Mannar, the southeast coast of India. *Environmental Geology, 45*, 466–480.

Laluraj, C. M., & Nair, S. M. (2006). Geochemical index of trace metals in the surficial sediments from the western continental shelf of India, Arabian Sea. *Environmental Geochemistry and Health, 28*(6), 509–518. https://doi.org/10.1007/s10653-005-8619-7

Li, H., Gao, X., Gu, Y., Wang, R., Xie, P., Liang, M., Ming, H., & Su, J. (2018). Comprehensive large-scale investigation and assessment of trace metal in the coastal sediments of Bohai Sea. *Marine Pollution Bulletin, 129*(1), 126–134. https://doi.org/10.1016/j.marpolbul.2018.02.022

Li, Q., Wu, Z., Chu, B., Zhang, N., Cai, S., & Fang, J. (2007). Heavy metals in coastal wetland sediments of the Pearl River Estuary, China. *Environmental Pollution, 149*(2), 158–164.

Monien, P., Lettmann, K. A., Monien, D., Asendorf, S., Woelfl, A. C., Lim, C. H., Thal, J., Schnetger, B., & Brumsack, H. J. (2014). Redox conditions and trace metal cycling in coastal sediments from the maritime Antarctic. *Geochimica et Cosmochimica Acta, 141*, 26–44. https://doi.org/10.1016/j.gca.2014.06.003

Preda, M., & Cox, M. E. (2002). Trace metal occurrence and distribution in sediments and mangroves, Pumicestone region, Southeast Queensland, Australia. *Environment International, 28*(5), 433–449. https://doi.org/10.1016/S0160-4120(02)00074-0

Santos, I. R., Silva-Filho, E. V., Schaefer, C. E. G. R., Albuquerque-Filho, M. R., & Campos, L. S. (2005). Heavy metal contamination in coastal sediments and soils near the Brazilian Antarctic Station, King George Island. *Marine Pollution Bulletin, 50*(2), 185–194. https://doi.org/10.1016/j.marpolbul.2004.10.009

Santos, I. R., Fávaro, D. I. T., Schaefer, C. E. G. R., & Silva-Filho, E. V. (2007). Sediment geochemistry in coastal maritime Antarctica (Admiralty Bay, King George Island): Evidence from rare earths and other elements. *Marine Chemistry, 107*(4), 464–474. https://doi.org/10.1016/j.marchem.2007.09.006

Siddique, A., Mumtaz, M., Zaigham, N. A., Mallick, K. A., Saied, S., Zahir, E., & Khwaja, H. A. (2009). Heavy metal toxicity levels in the coastal sediments of the Arabian Sea along the urban Karachi (Pakistan) region. *Marine Pollution Bulletin, 58*(9), 1406–1414. https://doi.org/10.1016/j.marpolbul.2009.06.010

Sun, W. P., Hu, C. Y., Weng, H. X., Han, Z. B., Shen, C., & Pan, J. M. (2013). Sources and geographic heterogeneity of trace metals in the: Sediments of Prydz Bay, East Antarctica. *Polar Research, 32*(SUPPL). https://doi.org/10.3402/polar.v32i0.20049

Trevizani, T. H., Figueira, R. C. L., Ribeiro, A. P., Theophilo, C. Y. S., Majer, A. P., Petti, M. A. V., Corbisier, T. N., & Montone, R. C. (2016). Bioaccumulation of heavy metals in marine organisms and sediments from Admiralty Bay, King George Island, Antarctica. *Marine Pollution Bulletin, 106*(1–2), 366–371. https://doi.org/10.1016/j.marpolbul.2016.02.056

Youssef, M., & El-Sorogy, A. (2016). Environmental assessment of heavy metal contamination in bottom sediments of Al-Kharrar lagoon, Rabigh, Red Sea, Saudi Arabia. *Arabian Journal of Geosciences, 9*(6), 1–10. https://doi.org/10.1007/s12517-016-2498-3

Zhao, G., Ye, S., Yuan, H., Ding, X., & Wang, J. (2017). Surface sediment properties and heavy metal pollution assessment in the Pearl River Estuary, China. *Environmental Science and Pollution Research, 24*(3), 2966–2979. https://doi.org/10.1007/s11356-016-8003-4

Zhu, A., Liu, J., Qiao, S., & Zhang, H. (2020). Distribution and assessment of heavy metals in surface sediments from the Bohai Sea of China. *Marine Pollution Bulletin, 153*(February), 110901. https://doi.org/10.1016/j.marpolbul.2020.110901

Chapter 10
Polycyclic Aromatic Hydrocarbons (PAHs) in the Aquatic Environment: A Rudimentary Approach

A. Ramzi, K. Habeeb Rahman, and Krishna K. Muraleedharan

10.1 Introduction

Polycyclic aromatic hydrocarbons (PAHs) are a diverse group of hydrophobic organic compounds with multiple, fused aromatic rings (Ankley et al., 1994; Law et al., 1997; Boese et al., 1999) that are linear, angular, or clustered type. Naphthalene ($C_{10}H_8$) is the member with the lowest molecular weight. They occur in oil, coal, or tar deposits and are by-products of fuel burning (whether fossil fuel or biomass). Out of over hundred PAHs and their structural isomers found in the environment, 16 of them are listed as priority pollutants by the USEPA. They are of concern because some of them have been identified as carcinogenic, mutagenic, and teratogenic.

PAHs are also found in cooked foods. Studies have shown that high levels of PAHs are found in meat exposed to high-temperature cooking, such as grilling or barbecuing, and in smoked fish. PAHs show variable physical-chemical and biological properties due to differences in position and number aromatic rings (Neff, 1979) (Fig. 10.1); they are generally classified into two groups based on their molecular weight (Law et al., 1997). Low molecular weight PAHs have high acute toxicity to aquatic species. However, most proven PAH carcinogens are among the high molecular weight congeners. Major PAHs with their molecular weights are listed in Table 10.1.

A. Ramzi (✉) · K. Habeeb Rahman · K. K. Muraleedharan
Department of Chemical Oceanography, Cochin University of Science and Technology, Kochi, India

H. Alshemmari et al. (eds.), *Contaminated Land and Water*, Emerging Contaminants and Associated Treatment Technologies, https://doi.org/10.1007/978-3-031-65129-8_10

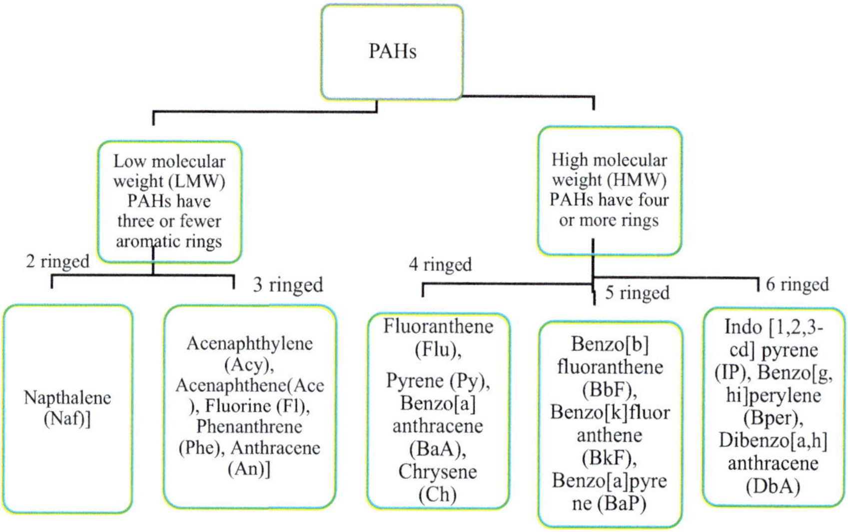

Fig. 10.1 Classification of PAHs based on number of rings

10.2 Sources

Combustion of fossil fuels release carbon dioxide and water; incomplete combustion results in the release of PAHs. Anthropogenic activities are their primary source to the environment. PAHs are broadly classified as pyrogenic and petrogenic origin (Wang et al., 2012; Keshavarzifard et al., 2015; Li et al., 2015). Polycyclic quinones are synthesized by bacteria, fungi, microalgae, etc. Under reducing conditions, these aromatic quinones are readily reduced to corresponding hydroquinones and subsequently to the parent PAHs (Aizenshtat 1973) (Fig. 10.2).

10.3 Solubility and Vapor Pressure

Vapor pressure and aqueous solubility of PAHs decrease with increasing molecular weight, influencing their distribution and behavior in the biological environment; thus, they are usually adsorbed on to the particulate matter in the atmosphere (Ravindra et al., 2008; Keshavarzifard et al., 2014).

They are generally insoluble in water but can be readily solubilized in organic solvents or organic acids. The most soluble PAH, naphthalene has a solubility of about 30 ppm, while five-ringed PAHs have solubilities in the 0.5–5.0 ppb range in distilled water. Alkyl-substituted PAHs tend to have relatively lower solubilities than the unsubstituted parent PAHs. Some of the exceptions include benz[a]anthracene which is less soluble than methyl or ethyl benz[a]anthracene and chrysene

Table 10.1 Major PAHs with their molecular weights and structural identity

IUPAC nomenclature	Molecular weight	Structural identity	Vapor pressure (Torr)	Solubility in water (mg L^{-1})
Naphthalene (Naf)	128.29	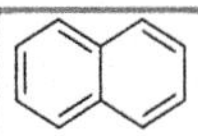	0.0492	32
Acenaphthylene (Acy)	152.21	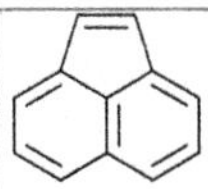	10^{-3}–10^{-2} at 20 °C	3.93
Acenaphthene (Ace)	154.21	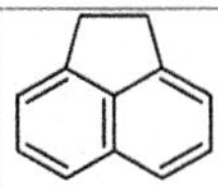	10^{-3}–10^{-2} at 20 °C	3.4 at 25 °C
Fluorene (Fl)	166.23	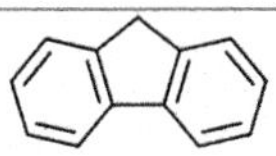	10^{-3}–10^{-2} at 20 °C	1.9
Phenanthrene (Phe)	178.24	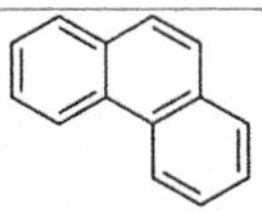	6.8×10^{-4} at 20 °C	1.0–1.3 at 25 °C
Anthracene (An)	178.24	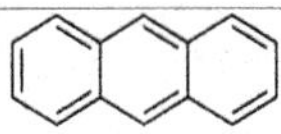	2×10^{-1} at 20 °C	0.05–0.07 at 25 °C
Fluoranthene (Flu)	202.26	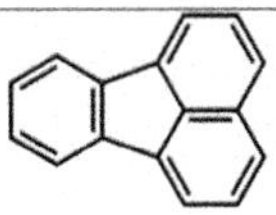	10^{-6}–10^{-4} at 20 °C	0.26 at 25 °C
Pyrene (Py)	202.26	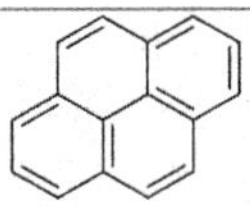	6.9×10^{-9} at 20 °C	0.14 at 25 °C
Benz(a)anthracene (BaA)	228.30	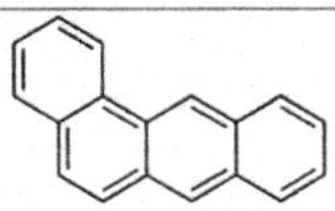	5×10^{-9} at 20 °C	0.01 at 25 °C
Chrysene (Ch)	228.30	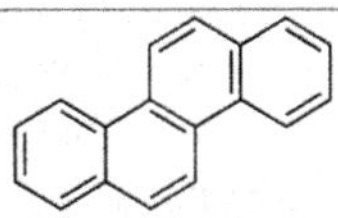	10^{-11}– 10^{-6} at 20 °C	0.002 at 25 °C
Benzo(b)fluoranthene (BbF)	252.32	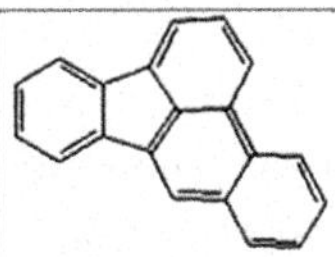	10^{-11}– 10^{-6} at 20 °C	–
Benzo(k)fluoranthene (BkF)	252.32	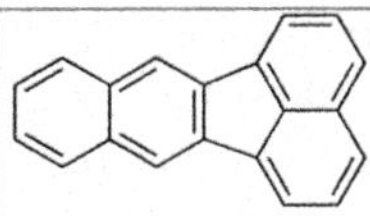	9.6×10^{-7} at 20 °C	–

(continued)

Table 10.1 (continued)

IUPAC nomenclature	Molecular weight	Structural identity	Vapor pressure (Torr)	Solubility in water (mg L^{-1})
Benzo(a)pyrene (BaP)	252.32	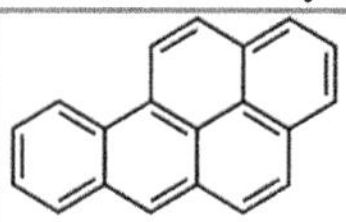	5×10^{-9} at 20 °C	0.01 at 25 °C
Benzo(ghi)perylene (BPer)	276.34	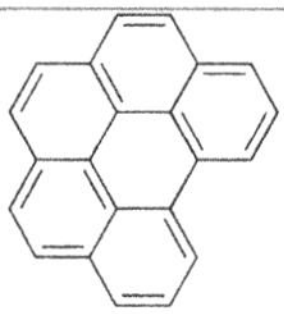	-10^{-10}	0.00026 at 25 °C
Inden(1,2,3-cd)pyrene (IP)	276.34	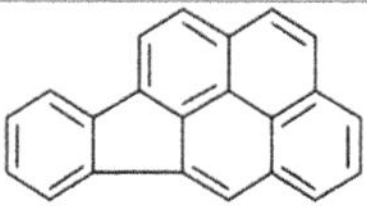	-10^{-10}	–
Dibenzo(ah) anthracene (DbA)	278.36	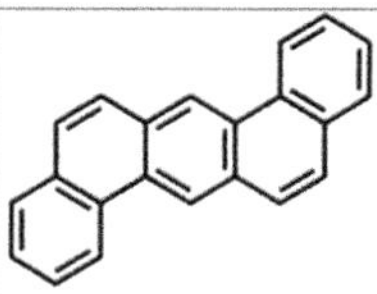	-10^{-10}	0.0005 at 25 °C

IUPAC International Union of Pure and Applied Chemistry

which is less soluble than methyl and dimethyl chrysene (Davis et al., 1942 cited by Neff, 1979). Linear PAHs are generally less soluble than the corresponding angular or pericondensed isomers. Thus, anthracene is less soluble than phenanthrene, and naphthacene is less soluble than chrysene, benz[a]anthracene, or triphenylene. Generally, PAHs are less soluble in seawater than in freshwater, a consequence of salting out. Solubilities of PAHs tend to increase with temperature. In natural waters, surfactants, such as humic and fulvic acids, and other surface-active substances may act as PAH solubilizers. In aqueous environments, therefore, PAHs are generally found adsorbed on particulates and on humic matter or solubilized in any oily matter which may contaminate water, sediments, and soil. The solubility of PAHs in water is inversely proportional to the number of constituent rings. As a result, larger molecular weight PAHs (≤ 4 rings) are almost exclusively bound to particulate matter, while lower molecular weight PAHs (up to three rings) can also be found dissolved in water. Since PAHs tend to have low vapor pressures, they are usually adsorbed on to particulate matter in the atmosphere. The vapor pressure of PAHs too is inversely proportional to the number of constituent rings. Larger molecular weight PAHs (≤ 4 rings) are mostly adsorbed onto particular matter in atmospheric samples, while lower molecular weight PAHs can be found both free in the atmosphere and bound to particulates.

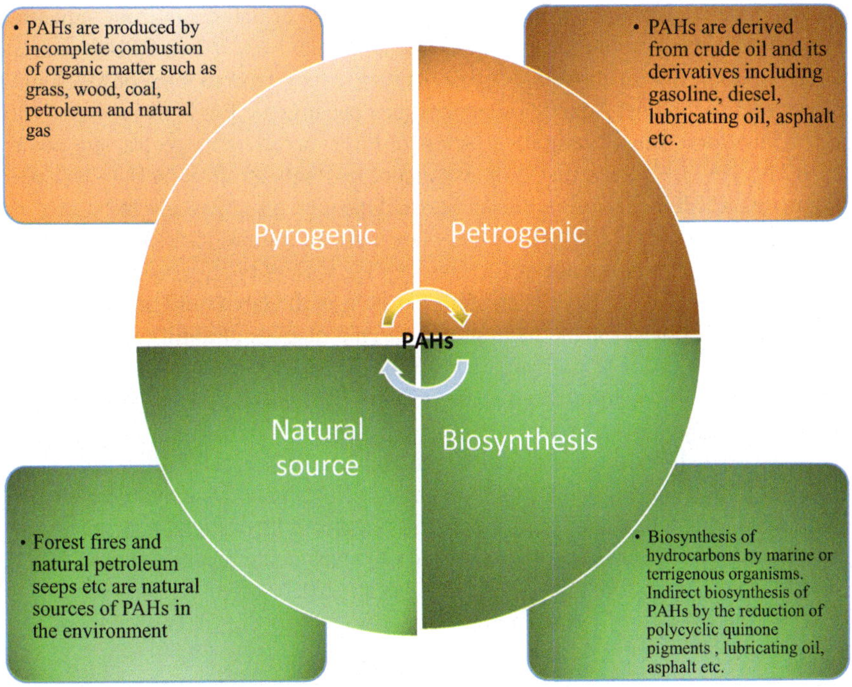

Fig. 10.2 Sources of PAHs into the environment

10.4 Distribution of PAHs in Water and Air-Water Interface

Once PAHs enter an aquatic system, a variety of physical and chemical processes, including biological interaction, influences their distribution in the water column. These processes include solubilization, partitioning between water and particulate matter, hydraulic transport, bioaccumulation, photodegradation and biodegradation, sedimentation and recycling, volatilization, and gas exchange at the air-water interface.

The air-water interface, especially the surface microlayer, plays a vital role in the transfer of substances between the atmosphere and the water column. In the atmosphere, low molecular weight (two to three rings) PAHs exist partially in the gaseous phase, whereas the relatively less volatile high molecular weight components prefer the particulate phase. Gaseous PAHs may be introduced into the sea surface through precipitation or by gaseous exchange at the air-water interface. Particulate PAHs may be deposited on the sea surface (and on land) in the wet form through precipitation. Particulates may also be deposited as dry fallout by gravitation supported by Brownian movement and trapping by whitecap bubbles on the sea surface.

The aquatic environment receives huge quantities of pollutants from industries, urban runoff, navigation, terrestrial discharge, mining, etc. When PAHs reach the water column, due to their high hydrophobicity, they tend to sorb on particulate matter and sediments. Aquatic organisms may accumulate PAHs from water, sediments, and food. Bioconcentration factors of PAHs in different species vary greatly (USEPA, 1993). Species that do not metabolize PAHs at all or only to a limited extent (algae, oligochaetes, mollusks, etc. and the more primitive invertebrates, protozoans, porifers, Cnidaria, etc.) accumulate high concentrations of PAHs.

Accumulation and release of PAHs by a wide variety of aquatic species, across different trophic levels, have been studied—both in the laboratory and in the field. Marine and freshwater algae have shown varying degrees of ability to accumulate and metabolize PAHs (e.g., Neff, 1979; Kriso & Irha, 1998). Kriso and Irha have reported 89–99% accumulation of benzo[a]pyrene in solution by marine brown algae *Fucus vesiculosus*.

However, fish and higher invertebrates that metabolize PAHs often accumulate little or no PAHs (arthropods, echinoderms, etc.). This may cause hydrocarbon contamination in sediment, and exposure to the benthic biota may lead to acute/chronic toxicity, whereas their transfer through the food web could result in serious human health effects (Weinstein et al., 2010; Zeng et al., 2013).

10.5 Degradation of PAHs

In the aquatic environment, degradation of PAHs occurs through abiotic and biotic processes. The most significant among them are photooxidation, chemical oxidation, and biodegradation. PAHs undergo photooxidation in the presence of sunlight, hydroxyl free radical, or other oxidizing agents to form endoperoxides, which on further photolysis or pyrolysis yield various oxidative products through ring cleavage and dealkylation (Neff, 1979). Some bacteria and yeast species are particularly efficient in degrading PAHs in marine and freshwater regions, except heavily polluted areas.

PAH exposure in humans is primarily through ingestion and inhalation. Under normal circumstances, skin contact with PAH is comparatively unimportant. Similar to other lipophilic compounds, PAHs usually are well absorbed in the body but are stored in the body very fewer, mainly in the kidney, liver, and spleen. Most of the absorbed amount is then expelled into bile and eventually feces and much lesser extent into urine. Most of the PAHs are excreted in a metabolized form, and only small amounts of the parent compound find their way into feces and urine (Neff, 1979). PAHs are highly soluble in fats. In this form, they can quickly enter the cells and become virtually unavailable for excretion. Metabolic processes tend to make PAHs more water-soluble, facilitating excretion.

10.6 Endocrine-Disrupting Property of PAHs

There are some ductless glands which secrete their hormones directly into the blood of vertebrates. These hormones reach particular parts of the body through blood circulation and induce specific actions. Some xenobiotic compounds can have hormone-like effects and can cause endocrine disruption (Ramzi, 2019). A compound that mimics the action of a particular hormone is called as agonist, whereas one that blocks its activity is termed antagonist. These compounds are called endocrine-disrupting chemicals (Lamb et al., 2014). The presence of endocrine disrupters (ECDs) can affect the endocrine system by many ways, such as the following:

1. Some chemicals can reduce the effects of natural hormones by blocking their interaction with their physiological binding sites.
2. They can react directly or indirectly with natural hormones.
3. They resemble the active site of hormones and can interact with receptor site.
4. They can alter the natural hormone synthesis pathway, etc. (Bergman et al., 2012).

The endocrine-disrupting activity is illustrated in Fig. 10.3.

Endocrine-disrupting chemicals are mostly man-made—pesticides, polychlorinated biphenyls, personal care products, phthalates, additives or contaminants in food, etc. ECDs are suspected to alter the reproductive system, inducing hormone-related cancer, thyroid disorders (such as abnormal growth in children), immune system disorders, metabolic disorders, bone disorders, etc. (Bergman et al., 2012). Increasing number of reports, over the past 50 years, about depressed sperm counts elevated incidents of breast cancer, and testicular cancer could be attributed to

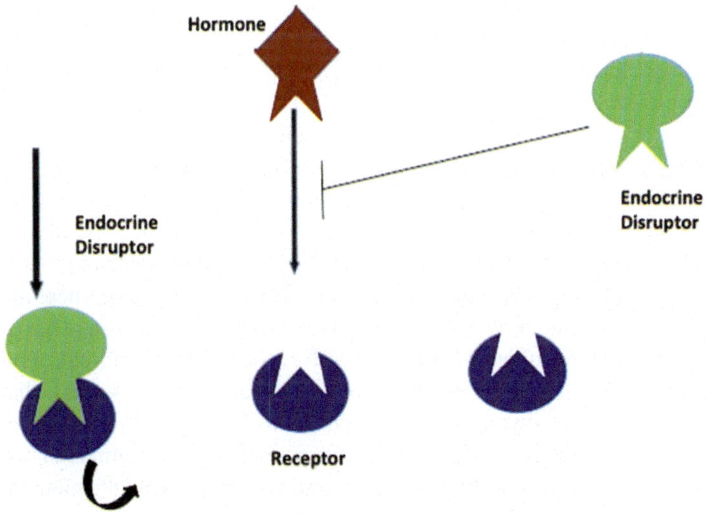

Fig. 10.3 Endocrine-disrupting activity of ECD

endocrine disruption. The study of background-level distribution and partitioning of these ECDs is very important.

10.7 Toxicity of PAHs

Metabolic processes of PAHs produce a variety of metabolites. Addition of one or more hydroxyl groups to the parent compound is a phase I reaction, whereas the addition of highly water-soluble groups to the PAH molecule is phase II reactions. Phase I reactions are regulated by epoxide enzymes hydrolase and a subset of cytochrome P-450 mixed-function oxidases called aryl hydrocarbon hydroxylase (AHH) (Vandermeulen & Penrose, 1978). The structures of PAH vary greatly, but the metabolism of these compounds is similar and leads to the formation of homologous metabolites (Neff, 1979; Bogovski et al., 1999; Yuan et al., 1999).

Due to the differences in the location of the metabolic modifications and the activities of the intermediate metabolites, they cause different tumors. Some of the metabolites are diol epoxide form; these diol epoxides are converted into carbonium ions. The carbonium ions can directly interact with DNA and proteins to form adducts and induce genotoxic damage (Neff, 1979). These alkylating agents are primary carcinogens, acting as initiators. Initiation is the first step in the development of cancer. The enzymes required for the conversion of parent PAH compounds into the reactive diol epoxides are found mainly in the liver but also the lungs, skin basal cell layer, intestinal mucosa, and other tissues (Neff, 1979).

PAHs may cause several toxic effects. Short-term exposure to a high dose of several PAHs has been shown to cause death in rodents. Since environmental levels are generally much lower than some of the occupational ones, it is extremely unlikely that short-term exposure to PAH in the environment would lead to death. On the other hand, eye irritation, photophobia, and skin toxicity such as dermatitis and keratosis have been demonstrated to be caused by occupational exposures to PAH (Boese et al., 1999).

Adverse respiratory effects, including acute and subacute inflammation and fibrosis, have been proven experimentally with B[a]Pyrine. Severe and long-lasting hyperplasia and metaplasia were observed as precancerous lesions and are consistent with the general assertion that one of the main targets of PAH toxicity is the respiratory tract (Lotufo, 1997). Carcinogenic PAHs (not the noncarcinogenic ones) have been reported to suppress immunity in rodents. Furthermore, there appears to be a rough correlation between the potency of PAH as immunosuppressors and as carcinogens. Immunosuppression may, therefore, be an important toxic endpoint of PAH (Ramzi, 2019; Payne & Fancey, 1989). Some authors also suggest that immunosuppression may be involved in the mechanisms by which PAH induce cancer.

Both in vivo tests in rodents and in vitro tests using mammalian (including human) cell lines, as well as in prokaryotes, were done to explain genotoxic effects for some PAHs (Yuan et al., 1999). Some PAHs appear not to be genotoxic but need to be metabolized first. Then the diol epoxides that are formed react with DNA to

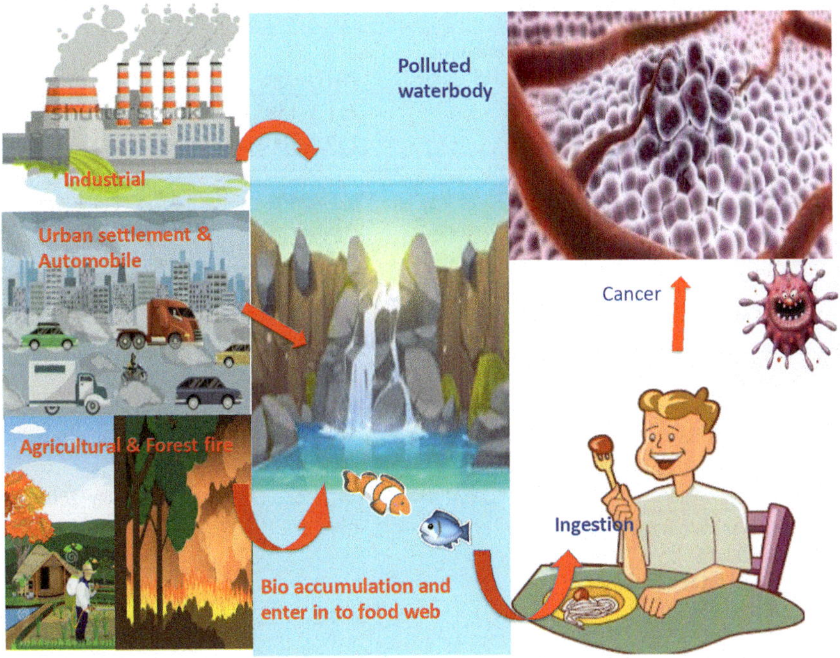

Fig. 10.4 PAH transport in the environment

form DNA adducts, thus inducing genotoxic damage. Effects like skin, eye, and respiratory mucosa irritation are more likely to be detected with severe exposures rather than with the characteristically lower environmental exposures (Yuan et al., 1999).

Persistent organic pollutants (POPs) have become the focal point of pollution research owing to their toxicity, carcinogenicity, and mutagenicity which might be harmful to human and/or environmental health (Wang et al., 2012; Keshavarzifard et al., 2015; Li et al., 2015). The USEPA has regulated at least 16 PAH fractions owing to their, persistence, prevalence, and potential hazards to the environment (USEPA, 2012) (Fig. 10.4).

The level of PAHs in water, particulate matter, and sediments reported from various areas in the world was tabulated below (Table 10.2).

10.8 Conclusion

The biosphere is under increasing anthropogenic influence of PAHs. Being persistent organic pollutants (POPs), they have become the focal point of pollution research for several decades owing to their toxicity, carcinogenicity, and mutagenicity which may be harmful to ecosystems and human health. People should be more concern about the use, management, and release of PAHs into the environment.

Table 10.2 PAHs reported worldwide

Locations Dissolved PAHs in (ng/L)	Number of PAHS studied	PAH range	References
Jialing River, China	16	811.5–1585.8	Xu et al. (2012)
Gulf of Tunis, Tunisia	24	139.2–1008.3	Mzoughi and Chouba (2011)
Bahia Blanca Estuary, Argentina Europe	17	n.d–4870	Arias et al. (2009)
Western Taiwan Strait, South China Sea	15	12.3–58.0	Wu et al. (2011)
Daliao River watershed, China	18	946.1–13448.5	Guo et al. (2007)
Gao-Ping River, Taiwan	16	10–9400	Doong and Lin (2004)
Tianjing, China	16	45.81–1272	Shi et al. (2005)
Hangzhou, China	10	989–9663	Chen et al. (2004)
Alexandria coast, Egypt	7	13–120	El Nemr and Abd-Allah (2003)
Daliao River Estuary, Bohai Sea South China	16	139.16–1717.87	Men et al. (2009)
Particulate PAHS in ng/g			
Gulf of Tunis, Tunisia	24	139.2–8222.4	Mzoughi and Chouba (2011)
Mundau´-Manguaba estuary, Brazil	16	454–1078	Maioli et al. (2011)
Daliao River Esturay, Bohai Sea, China	16	1542.44–20094.47	Men et al. (2009)
Daliao River watershed, China	18	317.5–238518.7	Guo et al. (2007)
Tianjing, China	16	938–64,200	Shi et al. (2005)
Pearl River Estuary, China	18	442–1850	Luo et al. (2006)
York River, VA Estuary, USA	20	199–1153	Countway et al. (2003)
Cochin estuary	16	60.02–60502.35	Ramzi et al. (2017)
Study area	**No. of PAHs Studied**	**PAHs (ng/g)**	**References**
Yangtze River Estuary, China	16	193–7588	Yu et al. (2015)
Pearl River estuary, China	16	126–3828	Zhang et al. (2015)
Luan River Estuary, China,	16	5–545	Zhang et al. (2015)
Yellow River Estuary, China	16	97.2–204.8	Hu et al. (2014)
Estuaries in Haihe River Basin, China	16	92–15,886	Liu et al. (2014)
Galician estuaries, NW Spain	35	44–7901	Perez-Fernandez et al. (2015)
Bahı ´a Blanca Estuary, Argentina	17	19–30,054	Oliva et al. (2015)
Selangor River Estuary, Malaysia	25	203–964	Masood et al. (2016)

(continued)

Table 10.2 (continued)

Locations Dissolved PAHs in (ng/L)	Number of PAHS studied	PAH range	References
Kaohsiung Harbor, Taiwan	17	34–16,700	Chen et al. (2013)
Mundaú-Manguaba estuarine-lagoon system, Brazil	16	29–222	Silva et al. (2013)
Patos Lagoon Estuary, Brazil	16	89–10,451	Garcia et al. (2010)
Esterode Urias, estuary, Mexico	11	27–418	Jaward et al. (2012)
Lenga Estuary, Chile	16	290–6118	Pozo et al. (2011)
Iko River estuary mangrove system, Nigeria	16	6100–35,270	Essien et al. (2011)
Hugli River, India	16	0–1839	Zuloaga et al. (2013)
Chitrapuzha River, India	16	5046–33,087	Kumar et al. (2016)
Cochin estuary, India	16	194–14,149	Ramzi et al. (2017)

References

Aizenshtat, Z. (1973). Perylene and its geochemical significance. *Geochimica et Cosmochimica Acta, 37*, 559–567.

Ankley, G. T., Thomas, N. A., Ditoro, D. M., Hansen, D. J., Mahony, J. D., Berry, W. J., Swartz, R. C., Hoke, R. A., Garrison, A. W., Allen, H. E., Zarba, C. S. (1994). Assessing potential bioavailability of metals in sediments - a proposed approach. *Environmental Management, 18*, 331–337.

Arias, A. H., Spetter, C. V., Freije, R. H., & Marcovecchio, J. E. (2009). Polycyclic aromatic hydrocarbons in water, mussels (Brachidontes sp., Tagelus sp.) and fish (Odontesthes sp.) from Bahía Blanca Estuary, Argentina. *Estuarine, Coastal and Shelf Science, 85*, 67–81.

Bergman, A., Heindel, J., Jobling, S., Kidd, K., & Zoeller, R. T. (2012). State-of-the-science of endocrine disrupting chemicals, 2012. *Toxicology Letters, 211*, S3–S3.

Boese, B. L., Ozretich, R. J., Lamberson, J. O., Swartz, R. C., Cole, F. A., Pelletier, J., & Jones, J. (1999). Toxicity and phototoxicity of mixtures of highly lipophilic PAH compounds in marine sediment: Can the Sigma PAH model be extrapolated? *Archives of Environmental Contamination and Toxicology, 36*, 270–280.

Bogovski, S., Sergeyev, B., Muzyka, V., & Karlova, S. (1999). Correlations between cytochrome P450, haem synthesis enzymes, and aromatic compounds in flounder (Platichthys flesus) from the Baltic Sea. *ICES Journal of Marine Science, 56*, 152–156.

Chen, B. L., Xuan, X. D., Zhu, L. Z., Wang, J., Gao, Y. Z., Yang, K., Shen, X. Y., & Lou, B. F. (2004). Distributions of polycyclic aromatic hydrocarbons in surface waters, sediments and soils of Hangzhou City, China. *Water Research, 38*, 3558–3568.

Chen, C.-F., Chen, C.-W., Dong, C.-D., & Kao, C.-M. (2013). Assessment of toxicity of polycyclic aromatic hydrocarbons in sediments of Kaohsiung Harbor, Taiwan. *Science of the Total Environment, 463*, 1174–1181.

Countway, R. E., Dickhut, R. M., & Canuel, E. A. (2003). Polycyclic aromatic hydrocarbon (PAH) distributions and associations with organic matter in surface waters of the York River, VA estuary. *Organic Geochemistry, 34*, 209–224.

Doong, R. A., & Lin, Y. T. (2004). Characterization and distribution of polycyclic aromatic hydrocarbon contaminations in surface sediment and water from Gao-Ping River, Taiwan. *Water Research, 38*, 1733–1744.

El Nemr, A., & Abd-Allah, A. M. A. (2003). Contamination of polycyclic aromatic hydrocarbons (PAHs) in microlayer and subsurface waters along Alexandria coast, Egypt. *Chemosphere, 52*, 1711–1716.

Essien, J. P., Eduok, S. I., & Olajire, A. A. (2011). Distribution and ecotoxicological significance of polycyclic aromatic hydrocarbons in sediments from Iko River estuary mangrove ecosystem. *Environmental Monitoring and Assessment, 176*, 99–107.

Garcia, M. R., Mirlean, N., Baisch, P. R., & Caramao, E. B. (2010). Assessment of polycyclic aromatic hydrocarbon influx and sediment contamination in an urbanized estuary. *Environmental Monitoring and Assessment, 168*, 269–276.

Guo, W., He, M. C., Yang, Z. F., Lin, C. Y., Quan, X. C., & Wang, H. Z. (2007). Distribution of polycyclic aromatic hydrocarbons in water, suspended particulate matter and sediment from Daliao River watershed, China. *Chemosphere, 68*, 93–104.

Hu, N.-J., Huang, P., Liu, J.-H., Ma, D.-Y., Shi, X.-F., Mao, J., & Liu, Y. (2014). Characterization and source apportionment of polycyclic aromatic hydrocarbons (PAHs) in sediments in the Yellow River estuary, China. *Environmental Earth Sciences, 71*, 873–883.

Jaward, F. M., Alegria, H. A., Galindo Reyes, J. G., & Hoare, A. (2012). Levels of PAHs in the waters, sediments, and shrimps of Estero de Urias, an Estuary in Mexico, and their toxicological effects. *The Scientific World Journal, 2012*, 1.

Keshavarzifard, M., Zakaria, M. P., Hwai, T. S., Yusuff, F. M., Mustafa, S., Vaezzadeh, V., Magam, S. M., Masood, N., Alkhadher, S. A. A., & Abootalebi-Jahromi, F. (2014). Baseline distributions and sources of Polycyclic Aromatic Hydrocarbons (PAHs) in the surface sediments from the Prai and Malacca Rivers, Peninsular Malaysia. *Marine Pollution Bulletin, 88*, 366–372.

Keshavarzifard, M., Zakaria, M. P., Hwai, T. S., Yusuff, F. M., & Mustafa, S. (2015). Distributions and source apportionment of sediment-associated polycyclic aromatic hydrocarbons (PAHs) and hopanes in rivers and estuaries of Peninsular Malaysia. *Environmental Science and Pollution Research, 22*, 9424–9437.

Kirso, U., & Irha, N. (1998). Role of algae in fate of carcinogenic polycyclic aromatic hydrocarbons in the aquatic environment. *Ecotoxicology and Environmental Safety, 41*, 83–89.

Kumar, K. S. S., Nair, S. M., Salas, P. M., Peter, K. J. P., & Kumar, C. S. R. (2016). Aliphatic and polycyclic aromatic hydrocarbon contamination in surface sediment of the Chitrapuzha River, South West India. *Chemistry and Ecology, 32*, 117–135.

Lamb, J. C., Boffetta, P., Foster, W. G., Goodman, J. E., Hentz, K. L., Rhomberg, L. R., Staveley, J., Swaen, G., Van Der Kraak, G., & Williams, A. L. (2014). Critical comments on the WHO-UNEP state of the science of endocrine disrupting Chemicals-2012. *Regulatory Toxicology and Pharmacology, 69*, 22–40.

Law, R. J., Dawes, V. J., Woodhead, R. J., & Matthiessen, P. (1997). Polycyclic aromatic hydrocarbons (PAH) in seawater around England and Wales. *Marine Pollution Bulletin, 34*, 306–322.

Li, J., Dong, H., Zhang, D., Han, B., Zhu, C., Liu, S., Liu, X., Ma, Q., & Li, X. (2015a). Sources and ecological risk assessment of PAHs in surface sediments from Bohai Sea and northern part of the Yellow Sea, China. *Marine Pollution Bulletin, 96*, 485–490.

Liu, F., Yang, Q., Hu, Y., Du, H., & Yuan, F. (2014). Distribution and transportation of polycyclic aromatic hydrocarbons (PAHs) at the Humen river mouth in the Pearl River delta and their influencing factors. *Marine Pollution Bulletin, 84*, 401–410.

Lotufo, G. R. (1997). Toxicity of sediment-associated PAHs to an estuarine copepod: Effects on survival, feeding, reproduction and behavior. *Marine Environmental Research, 44*, 149–166.

Luo, X. J., Chen, S. J., Mai, B. X., Yang, Q. S., Sheng, G. Y., & Fu, J. M. (2006). Polycyclic aromatic hydrocarbons in suspended particulate matter and sediments from the Pearl River Estuary and adjacent coastal areas, China. *Environmental Pollution, 139*, 9–20.

Maioli, O. L. G., Rodrigues, K. C., Knoppers, B. A., & Azevedo, D. A. (2011). Distribution and sources of aliphatic and polycyclic aromatic hydrocarbons in suspended particulate matter in water from two Brazilian estuarine systems. *Continental Shelf Research, 31*, 1116–1127.

Masood, N., Zakaria, M. P., Halimoon, N., Aris, A. Z., Magam, S. M., Kannan, N., Mustafa, S., Ali, M. M., Keshavarzifard, M., Vaezzadeh, V., Alkhadher, S. A. A., & Al-Odaini, N. A. (2016). Anthropogenic waste indicators (AWIs), particularly PAHs and LABs, in Malaysian sediments: Application of aquatic environment for identifying anthropogenic pollution. *Marine Pollution Bulletin, 102*, 160–175.

Men, B., He, M. C., Tan, L., Lin, C. Y., & Quan, X. C. (2009). Distributions of polycyclic aromatic hydrocarbons in the Daliao River Estuary of Liaodong Bay, Bohai Sea (China). *Marine Pollution Bulletin, 58*, 818–826.

Mzoughi, N., & Chouba, L. (2011). Distribution and partitioning of aliphatic hydrocarbons and polycyclic aromatic hydrocarbons between water, suspended particulate matter, and sediment in harbours of the west coastal of the Gulf of Tunis (Tunisia). *Journal of Environmental Monitoring, 13*, 689–698.

Neff, J. M. (1979). *Polycyclic aromatic hydrocarbons in the aquatic environment – Sources, fates and biological effects* (p. 262). Applied Science Publishers.

Oliva, A. L., Quintas, P. Y., La Colla, N. S., Arias, A. H., & Marcovecchio, J. E. (2015). Distribution, sources, and potential Ecotoxicological risk of polycyclic aromatic hydrocarbons in surface sediments from Bahia Blanca Estuary, Argentina. *Archives of Environmental Contamination and Toxicology, 69*, 163–172.

Payne, J. F., & Fancey, L. F. (1989). Effect of polycyclic aromatic-hydrocarbons on immune-responses in fish – Change in melanomacrophage centers in flounder (pseudopleuronectes-americanus) exposed to hydrocarbon-contaminated sediments. *Marine Environmental Research, 28*, 431–435.

Perez-Fernandez, B., Vinas, L., Franco, M. A., & Bargiela, J. (2015). PAHs in the ria de Arousa (NW Spain): A consideration of PAHs sources and abundance. *Marine Pollution Bulletin, 95*, 155–165.

Pozo, K., Perra, G., Menchi, V., Urrutia, R., Parra, O., Rudolph, A., & Focardi, S. (2011). Levels and spatial distribution of polycyclic aromatic hydrocarbons (PAHs) in sediments from Lenga Estuary, Central Chile. *Marine Pollution Bulletin, 62*, 1572–1576.

Ramzi, A. (2019). *Dynamics and ecological risk of endocrine disrupting organic pollutants in Cochin Estuary.*

Ramzi, A., Rahman, K. H., Gireeshkumar, T. R., Balachandran, K. K., Chacko, J., & Chandramohanakumar, N. (2017). Dynamics of polycyclic aromatic hydrocarbons (PAHs) in surface sediments of Cochin Estuary, India. *Marine Pollution Bulletin, 114*, 1081–1087.

Ravindra, K., Sokhi, R., & Van Grieken, R. (2008). Atmospheric polycyclic aromatic hydrocarbons: Source attribution, emission factors, and regulation. *Atmospheric Environment, 42*(13), 2895–2921.

Shi, Z., Tao, S., Pan, B., Fan, W., He, X. C., Zuo, Q., Wu, S. P., Li, B. G., Cao, J., Liu, W. X., Xu, F. L., Wang, X. J., Shen, W. R., & Wong, P. K. (2005). Contamination of rivers in Tianjin, China by polycyclic aromatic hydrocarbons. *Environmental Pollution, 134*, 97–111.

Silva, T. R., Lopes, S. R. P., Spoerl, G., Knoppers, B. A., & Azevedo, D. A. (2013). Evaluation of anthropogenic inputs of hydrocarbons in sediment cores from a tropical Brazilian estuarine system. *Microchemical Journal, 109*, 178–188.

U.S.E.PA (Environmental Protection Agency). (2012). *Sediment contamination.*

USEPA. (1993). *Provisional guidance for quantitative risk assessment of polycyclic aromatic hydrocarbons*. EPA/600/R-93/089. Office of Research and Development.

Vandermeulen, J. H., & Penrose, W. R. (1978). Absence of aryl hydrocarbon hydroxytase (AHH) in three marine bivalves. *Journal of the Fisheries Research Board of Canada, 35*, 643–647.

Wang, Y., Li, X., Li, B. H., Shen, Z. Y., Feng, C. H., & Chen, Y. X. (2012). Characterization, sources, and potential risk assessment of PAHs in surface sediments from nearshore and farther shore zones of the Yangtze estuary, China. *Environmental Science and Pollution Research, 19*, 4148–4158.

Weinstein, J. E., Crawford, K. D., Garner, T. R., & Flemming, A. J. (2010). Screening-level ecological and human health risk assessment of polycyclic aromatic hydrocarbons in stormwater detention pond sediments of coastal South Carolina, USA. *Journal of Hazardous Materials, 178*, 906–916.

Wu, Y. L., Wang, X. H., Li, Y. Y., & Hong, H. S. (2011). Occurrence of polycyclic aromatic hydrocarbons (PAHs) in seawater from the Western Taiwan Strait, China. *Marine Pollution Bulletin, 63*, 459–463.

Xu, X. Y., Jiang, Z. Y., Wang, J. H., Hong, Z., Dou, J. F., Narres, H. D., Hofmann, D., & Klumpp, E. (2012). Distribution and characterizing sources of polycyclic aromatic hydrocarbons of surface water from Jialing River. *Journal of Central South University of Technology (English Edition), 19*, 850–854.

Yu, W., Liu, R., Xu, F., & Shen, Z. (2015). Environmental risk assessments and spatial variations of polycyclic aromatic hydrocarbons in surface sediments in Yangtze River estuary, China. *Marine Pollution Bulletin, 100*, 507–515.

Yuan, Z. X., Honey, S. A., Kumar, S., & Sikka, H. C. (1999). Comparative metabolism of dibenzo a,l pyrene by liver microsomes from rainbow trout and rats. *Aquatic Toxicology, 45*, 1–8.

Zeng, S., Zeng, L., Dong, X., & Chen, J. (2013). Polycyclic aromatic hydrocarbons in river sediments from the western and southern catchments of the Bohai Sea, China: Toxicity assessment and source identification. *Environmental Monitoring and Assessment, 185*, 4291–4303.

Zhang, J.-D., Wang, Y.-S., Cheng, H., Jiang, Z.-Y., Sun, C.-C., & Wu, M.-L. (2015). Distribution and sources of the polycyclic aromatic hydrocarbons in the sediments of the Pearl River estuary, China. *Ecotoxicology, 24*, 1643–1649.

Zuloaga, O., Prieto, A., Ahmed, K., Sarkar, S. K., Bhattacharya, A., Chatterjee, M., Bhattacharya, B. D., & Satpathy, K. K. (2013). Distribution of polycyclic aromatic hydrocarbons in recent sediments of Sundarban mangrove wetland of India and Bangladesh: A comparative approach. *Environmental Earth Sciences, 68*, 355–367.

Chapter 11
Recent Advances in Degradation of Textile Dyes in Wastewater Under UV Visible Wavelength by Nanocomposite: A Review

Jabran Ahmad, Muhammad Zaffar Hashmi, Aamer Saeed, Siwatt Pongpiachan, Xiaomei Su, Arianit A. Reka, and Zulkfil Ahmed

11.1 Introduction

A highly comfortable lifestyle developed by human beings has a great polluted impact on the environment (Smrithi et al., 2020). In the last 20 years, supply of pure water is a big problem in developing countries that concerns with health issues. In a published report of the World Health Organization (WHO) in 2014, it showed that over 10 lac people embedded in serious health issues due to water contamination (Gnanasekaran et al., 2017). In recent times, textile dyes present in industrial wastewater cause a serious threat for a stable ecosystem and human health (Wang et al., 2017). Dyes are used in great amount in textile, paints, and paper industry. However, the presence of textile dyes in wastewater is a big problem for environment and

J. Ahmad
Department of Chemistry, COMSATS University, Islamabad, Pakistan

M. Z. Hashmi (✉)
Institute of Molecular biology and Biotechnology, The University of Lahore, Lahore, Pakistan

A. Saeed
Department of Chemistry, Quaid-i-Azam University Islamabad, Islamabad, Pakistan

S. Pongpiachan
School of Social and Environmental Development, National Institute of Development Administration (NIDA), Bangkok, Thailand

X. Su
Department of Environmental Sciences, Zhejiang Normal University, Jinhua, China

A. A. Reka
Faculty of Natural Sciences and Mathematics, Department of Chemistry, University of Tetovo, Tetovo, Republic of North Macedonia

Z. Ahmed
Mir Chakar Khan Rind University, Dera Ghazi Khan, Pakistan

H. Alshemmari et al. (eds.), *Contaminated Land and Water*, Emerging Contaminants and Associated Treatment Technologies, https://doi.org/10.1007/978-3-031-65129-8_11

human health (Alshabanat & AL-Anazy, 2018). Pure water resources are necessary for continuation of healthy and safe life. But the revolution in industry and urbanization make it impossible in the twenty-first century. These increase many organic and inorganic dyes in wastewater of textile effluents (Taymaz et al., 2021). Textile dyes may be used in small quantity, but due to its water organoleptic properties, it blocks the sunlight to penetrate inside and hence disturb the aquatic life. On other hand, some dyes due to their mutagenic and cancerous affect make water unable to consume (Martins et al.).

Norio Taniguchi (a professor in the University of Science, Tokyo) first invented the term of nanoparticles. Nano is a word of Greek language which means very small (Nagajyothi et al., 2020). Nanoparticles have one or more external dimensions having particle size 1–100 nm. Due to small size of nanoparticles, they have remarkable properties, i.e., large surface area, more active sites, high crystallinity, quantum effect, and unique electron conduction property. On the basis of these properties, nanoparticles are greatly used as adsorbents, catalysts, and sensors and in other applications. The nanoparticles show superior performance than bulk materials (Cai et al., 2017). In modern age, nanoparticles have a wide range of applications in different fields. Herein, we study the application of nanoparticles in the synthesis of nanocomposites for the removal of textile dyes, i.e., $BiVO_4$/rGO is also a nanocomposite used for the degradation of methylene blue and Congo red dyes with removal efficiency of 95% and 98%, respectively, by photocatalytic degradation (Appavu et al., 2018). Nanoparticles are employed with different metal oxide to synthesize a metal oxide composite. The purpose of metal oxide composite is their application in the elimination of toxic dyes from wastewater. UV/visible is a spectroscopic technique and known as the highly simplified and convenient method for analyzing the particle interaction with light. This employs as a qualitative method (Bumbrah & Sharma, 2016). The variance in absorption is analyzed as a function of wavelength. It is known as a versatile and fast building result technique (Mujahid & Dickert, 2012). UV/visible spectroscopy is one of the highly significant characterization techniques to study optical properties of mattes. It assists in the understanding of interaction of matter with light. This technique is mostly applied for the determination of concentration of solutes. It is a type of absorption spectroscopy (Pentassuglia et al., 2018). UV/Visible spectroscopy base on Lambert-Beer Law. The mathematical equation for quantitative measurement is as follows (Wang & Chu, 2013):

$$\mathrm{A} = \log_{10}\left(\mathrm{I}_0 / \mathrm{I}\right) = \varepsilon c l$$

Dyes are defined as the colored material that is used for coloring the substances. Dyes can be natural (animal, vegetable, and mineral sources) or synthetic. Dyes are usually complex organic compounds that consist of two main components, namely, chromophores and auxochromes (Hasanpour & Hatami, 2020). Different industries used synthetic dyes like leather, paper, and more significantly textile. Textile industry is mainly known as the most highly utilizing body of dyes (Katheresan et al., 2018).

All types of dyes that are used in industries are toxic, carcinogenic, and a big threat for ecosystem (Kaviyarasu et al., 2020). The dyes are considered as a main factor of water pollution. The major portion of dyes comes through textile industry (Agustina, 2020). The high stability and mutagenicity of dyes are a big threat for humans and aquatic life (Oliveira et al., 2019). In the recent years, the environment of the world changes dramatically. The industrial revolution has a big impact on this environmental change (Wu et al., 2019). Therefore, treatment of dyes is necessary for the sustainable environment. The major object of this study is to illustrate the role of nanoparticles with metal oxide in the degradation of textile dyes.

11.2 Treatment Technologies

There are different methods to degrade dyes from textile effluents. These methods include physicochemical expansion, such as adsorption, adsorption on stimulated carbon, membrane separation, and reverse osmosis (RO). The drawback of these methods are that they are cost-effective, having large volume of wastage, and these methods also use some toxic additives that cause damage in the ecosystem (Magdalane et al., 2017). Biological methods are also employed to degrade textile dyes. The major drawback of this method is that some dyes are not biodegradable due to the complex structure and high molecular masses of dyes (Arcanjo et al., 2018). Most dyes are known as very stable molecule, and removing it from wastewater and making it toxic-free are considered difficult tasks. The use of dyes became necessary due to the demand of industries associated with it (Dutta et al., 2018). Photocatalysis also plays a vital role in the degradation of textile dyes from wastewater. Like other methods, this method also has limitation such as the removal of catalyst at the end of reaction with treated elute. On the other hand, the removal efficiency becomes remarkably low when the color and turbidity of wastewater are high (Kumaran et al., 2021). Recently, researchers are interested to develop a method to use nanocomposites in the degradation of dyes. Nowadays, semiconductor-based nanocomposite of metal oxide is widely used in the degradation of textile dyes due to their excellent efficiency toward the removal of toxic dyes (Gupta et al., 2017). The efficiency of semiconductor-based nanocomposite of metal oxide may be increased by heterogeneous photocatalysis. This method has a leading role toward the elimination of textile dyes due to it being highly efficient, low cost, and eco-friendly. The nanocomposite of metal oxide performs exceptionally well in the removal of dyes due to their high crystallinity, large surface area, and small size (Oliveira et al., 2019). Metal oxide semiconductor photocatalysts are used for water purification. The main advantage to use these metal oxides is that they do not generate any secondary pollutant.

11.3 Nanoparticles

Nanoparticles are defined as the solid particles or division of particles that have size in the range of 10–100 nm. Nanoparticles, nanospheres, nanorods, and nanocapsules can be obtained by using different methods of preparation (Mohanraj & Chen, 2006). Nanoparticles got importance when researchers found that size had great effect on physiochemical properties of a substance. The nanoparticles displayed many colors and characteristic properties with the change of size. Due to these characteristics, nanoparticles play an important role in different fields (Khan et al., 2019).

Nanoparticles are composed of organic and inorganic substances that have high capacity to adsorb toxic material due to their large surface area and small particle size (Jawaid et al., 2019). In the past, the development of simple, low-cost, and high-yield composition of nanoparticles was a big pebble for the scientist. There are many bottoms and top-down methods (i.e., physical and chemical) that were employed for the production of nanoparticles. In all top-down approaches, the most widely used method is high-energy ball milling. In this process, a sufficient powder charge is kept in a high-energy mill, by using an appropriate milling medium. The main purpose of this milling is to make the particle size small and convert it to a nanoparticle (Yadav et al., 2012). Liquid-phase laser ablation method was also a process applied for the preparation of nanoparticles. This procedure attracted great interest due to simplicity and versatility. This method was mainly used in the synthesis of Cu nanoparticles due to their difficult preservation in water (Saito et al., 2008). The use of sputtering dispersion technique was also applied for the production of nanoparticles. By using this technique, nanoparticles can be synthesized under mild condition by easily controlling their size and morphology with suitable tuning of sputtering conditions. This technique allows fast synthesis of metal nanoparticles (M-Nps) (Luza et al., 2013). Lithography is also a method for the synthesis of nanoparticles. Two types of lithography, namely, photolithography and scanning beam lithography, were employed for the nanoparticle's synthesis. This technique had drawbacks of high cost and exposure of substrate to high-energy radiations (Acikgoz et al., 2011). Another method known as photochemical reduction was also employed for the production of nanoparticles. This method was known as more convenient and eco-friendlier than other methods. Ag nanostructures, i.e., nanorods and nanowires, were developed by this method demonstrated with citrate-capped Pt seeds (Park et al., 2011). Thermal decomposition method was also a well-known method for the preparation of nanoparticles. The process consists of dissolution of protective stabilizer at an elevated boiling solvent, and then the necessary precursor metal is added to the solution and reflux which yield nanoparticles. This method was employed to synthesize tungsten nanoparticles (Sahoo et al., 2009). Reverse microemulsion is also a developing method for the production of nanoparticles. This method is used for the synthesize of nanoparticles from considering silica, oxides, polymers, semiconductors, and superconductors (Tojo et al., 2011).

Nowadays, biosynthesis is a replacement of physical and chemical methods for the preparation of nanoparticles. Different microbes such as bacteria, fungi, yeast, and viruses are used to yield nanoparticles (Nagajyothi et al., 2020). Biological processes to employ the production of nanoparticles are known as eco-friendly nano-factories. Platinum nanocomposites were synthesized by using microorganism, i.e., cyanobacterium *Plectonema boryanum*, the most vital group of bacteria species (Lengke et al., 2006). Anyhow, the application of microorganism to synthesize nanoparticles has major drawbacks such as slow process of synthesis, different procedure of synthesis, time-consuming, and low number of sizes and shapes. These limitations were overcome by employing the plant extracts for the synthesis of nanoparticles. This method is low cost, eco-friendly, and single step. The process mainly consist of two steps, i.e., choice of solvent and selection of environmentally friendly reducing and oxidizing agent and nontoxic substances for nanoparticle stability (Iravani et al., 2014). In modern age, nanoparticles have a wide range of applications in different fields. Herein, we study the application of nanoparticles in the synthesis of nanocomposites for the removal of textile dyes.

11.4 Metal Oxides

Metal oxides are defined as the crystalline solid that consists of a metal cation and an oxide anion. Metal oxides have a huge range of applications. Herein, we illustrate an overview of metal oxide to explore their role in the treatment of wastewater by degrading the textile dyes. This process can be summarized in a flow chart diagram as (Fig. 11.1):

TiO_2 is a metal oxide that is used in the elimination of textile dyes. Researchers reported degradation of different dyes by using TiO_2 metal oxide. In a reported

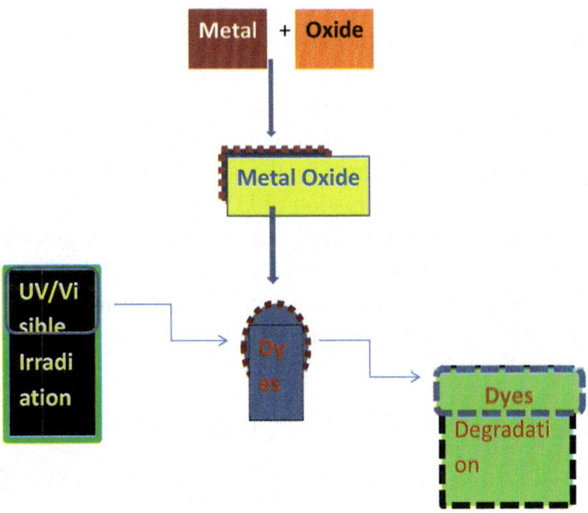

Fig. 11.1 Flow chart diagram of degradation of textile dyes by metal oxide under UV/visible wavelength

method, TiO_2 was applied for the removal of methylene blue, methyl violet, and acid violet. The degradation of methyl blue took place in 150 min, while other two dyes degraded in 6–7 h. The removal efficiency of methyl blue was 99%, while the other two were 72% (Rajagopal et al., 2020). In another reported method, TiO_2 was used for the elimination of malachite green (MG) and methylene blue (MB). The degradation time of both dyes was 5 hrs. The removal efficiency was also significant (Bansal et al., 2020). TiO_2 is the most widely used metal oxide for the degradation of dyes. A researcher employed TiO_2 for the removal of acid blue 113(AB 113) dye. The results of the degradation were exceptionally well. TiO_2 removed AB 113 from wastewater at wavelength lower than 190 nm with the efficiency of 98.7% (Mortazavian et al., 2019). A method was reported in which TiO_2 was utilized in the removal of azo dyes. The removal efficiency of TiO_2 toward degradation of azo dyes was 96% within time range of 6 h (Arcanjo et al., 2018). A researcher reported a method in which TiO_2 metal oxide was applied for the degradation of methylene blue, methyl orange, indigo carmine, and Chicago Sky Blue 6B. The degradation time of these dyes by TiO_2 metal oxide was 5 h. The removal efficiency reported by the researcher was 90.3%, 98.5%, 92.4%, and 60.3%, respectively (Zainal et al., 2005). On the basis of abovementioned results, it can be deducted that TiO_2 is an excellent metal oxide for the elimination of methylene blue dye.

$Sm_2Ti_2O_7$ is a binary metal oxide that also plays vital role in the degradation of textile dyes. In a reported method, $Sm_2Ti_2O_7$ was employed for the elimination of rhodamine B (RhB) dye. The degradation efficiency was reported to be 94% (Kaviyarasu et al., 2020). Nb_2O_5 is another metal oxide that also has application in the elimination of textile dyes from wastewater. A researcher reported a method in which Nb_2O_5 was used in the degradation of azo dyes. The results mentioned that Nb_2O_5 had excellent efficiency in elimination of azo dyes. The degradation took place in 40 min with 96% efficiency (Dos Santos et al., 2019). Co_3O_4 is a metal oxide that is also applied for the elimination of textile dyes from wastewater. In a reported method, Co_3O_4 was applied for the degradation of rhodamine B (RhB) dye. The reported results suggested that Co_3O_4 degraded dye in 30 min with 78% efficiency (Hitkari et al., 2018). SnO_2 is the most widely used metal oxide in the removal of textile dyes. In a reported method, a researcher applied SnO_2 in the degradation of methyl orange and rhodamine B. The results suggested that degradation took place in 100 min with 61% efficiency (Gupta et al., 2017). WO_3 is also a well-known metal oxide for the degradation of dyes. In a reported method, WO_3 was applied for the elimination of Orange II and 2-nitrophenol dyes (Table 11.1). The reported results showed that degradation took place in 120 and 150 min, respectively, with the removal efficiencies of 95% and 82% (Umukoro et al., 2017). On the basis of reported results, it could be suggested that $Sm_2Ti_2O_7$ is an excellent metal oxide for the removal of rhodamine B dye.

Table 11.1 Degradation of dyes under different UV/visible wavelength with removal efficiency regarding time by using metal oxide

Sr. no.	Metal oxides	Wastewater dyes	Wavelength (nm)	Time	Removal efficiency	References
1	$Sm_2Ti_2O_7$	Rhodamine B	498	–	94%	Kaviyarasu et al. (2020)
2	TiO_2	Methylene blue Methyl violet Acid violet	665 590 540	150 min 6–7 h 6–7 h	99% 72% 72%	Rajagopal et al. (2020)
3	TiO_2	Malachite green (MG), Methylene blue (MB)	528	5 h	–	Bansal et al. (2020)
4	TiO_2	AB 113	<190	–	98.7	Mortazavian et al. (2019)
5	TiO_2	Azo dyes	–	6 h	96%	Arcanjo et al. (2018)
6	TiO_2	Methylene blue, methyl orange indigo carmine Chicago sky blue 6B	661 505 466 618	4 h	90.3% 98.5% 92.4% 60.3%	Zainal et al. (2005)
7	Nb_2O_5	Azo dyes	–	40 min	96%	Dos Santos et al. (2019)
8	Co_3O_4	Rhodamine	–	30 min	78%	Hitkari et al. (2018)
9	SnO_2	Methyl orange Rhodamine B	365	100 min	61%	Gupta et al. (2017)
10	WO_3	Orange II 2-nitrophenol	480	120 min 180 min	95% 82%	Umukoro et al. (2017)

11.5 Metal Oxide Nanocomposites

Nanoparticles of metals are employed with different metal oxide to synthesize a polymer composite. These nanocomposites of metal oxides are used for the elimination of textile dyes from wastewater. This can be illustrated in the following diagram (Fig. 11.2):

TiO_2 has an excellent property in the elimination of textile dyes. The capability of TiO_2 adsorption is increased by the synthesis of TiO_2-graphene oxide composite formation. The nanocomposite shows even better properties than TiO_2 toward the removal of textile dyes. The nanocomposites are used to degrade methylene blue dye with 99% removal efficiency (Banerjee et al., 2018). This method is also used to synthesize nanocomposite of TiO_2 with AgO/Ag_2O nanoparticles. The nanocomposite used in the treatment of reactive blue 220 dye from the textile wastewater with removal efficiency of 96% (Kulal & Kodialbail, 2021). A method was reported in which TiO_2 was combined with Fe_3O_4 and SiO_2 to synthesize a polymer nanocomposite. The polymer nanocomposite $Fe_3O_4@TiO_2/SiO_2$ had excellent magnetic properties in the adsorption of textile dyes. The $Fe_3O_4@TiO_2/SiO_2$ polymer nanocomposite was applied for the elimination of rhodamine B dyes. The reported

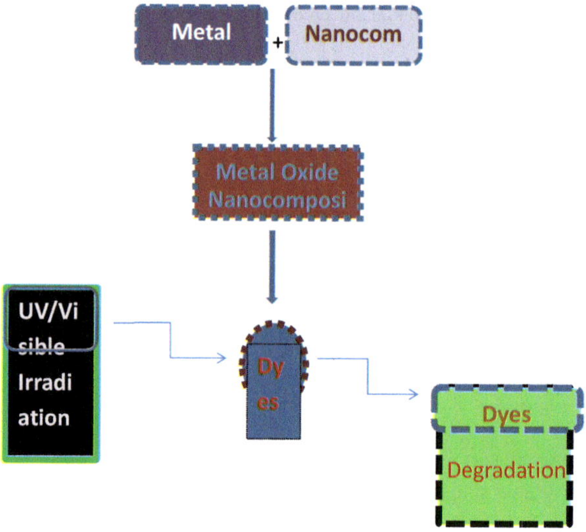

Fig. 11.2 Flow chart diagram of degradation of textile dyes by metal oxide nanocomposite under UV/visible wavelength

removal efficiency of the polymer composite was 97% (Li et al., 2016). A method was reported in which the Nd base TiO_2 nanocomposite was synthesized on graphene oxide with sol-gel process. The Nd-TiO_2-GO nanocomposite was employed for the removal of indigo carmine dye. The reported elimination efficiency was 92% (Oppong et al., 2016). In another approach, a nanocomposite was synthesized by using P_{25} with metal oxide TiO_2. P_{25}-TiO_2 nanocomposite had high rate of photocatalytic performance than pure TiO_2. P_{25}-TiO_2 nanocomposite was applied for the removal of rhodamine B dye. The nanocomposite of P_{25}-TiO_2 had exceptionally high rate of degradation of rhodamine B dye with the reported removal efficiency of 99% (Qin et al., 2009). Another method was reported in which TiO_2-S nanocomposite was synthesized. This nanocomposite was used for the elimination of methyl orange. The reported result suggested that degradation took place with 92.8% efficiency in 6 h (Yu et al., 2017). In another approach, a nanocomposite was synthesized by using TiO_2 nanosheets and CdSe nanocomposite. The synthesized nanocomposite had its application in the removal of methylene blue dye. The reported results suggested that TiO_2-CdSe nanocomposite removed methylene blue with 98.5% efficiency (Xu et al., 2019). A researcher reported a method in which the synthesis of Nd-TiO_2 was reported. The synthesized nanocomposite is applied for the degradation of indigo carmine. The results showed that in 3 h 92% degradation took place (Oppong et al., 2016). On the basis of above-reported result, it can be concluded that TiO_2-graphene oxide composite shows best degradation efficiency of textile dyes from wastewater.

Zn nanoparticles were used with many other metal oxides for the preparation of nanocomposites. The composition of Zn-TiO_2 nanoparticles demonstrates remarkable performance in photocatalytic elimination of dyes. The nanocomposite of

Zn-TiO_2 is used in the removal of azo dyes. The efficiency of eradication of dyes is 87–99% (Foulady-Dehaghi & Behpour, 2020). ZnO is one of the most widely used metal oxides in the degradation of textile dyes. The photodegradation of ZnO is highly evaluated by using Cu nanoparticles. The Cu-ZnO-doped nanocomposite yields 92% efficiency in the removal of navy blue dye (Kumaran et al., 2021). Another method to synthesize nanocomposite also employs ZnO nanoparticles. In this technique, neodymium was doped with ZnO to synthesize the nanocomposite. The synthesized nanocomposite was used to degrade the acid blue and eosin Y dyes. The nanocomposite yields 99% and 80% degradation efficiency, respectively (Anku et al., 2020). Rare earth metals are also used with metal oxide in the synthesis of nanocomposite for removal of dyes. In a reported method, rare earth metal, i.e., Gd (gadolinium), was doped with ZnO to evaluate the photocatalytic activity of ZnO toward the removal of textile dyes. The nanocomposite of Gd-ZnO degrades the methylene blue. The removal efficiency of nanocomposite was 93% (Selvaraj et al., 2019). The ZnO nanoparticles were also doped with light metal to form nanocomposites used in the degradation of textile dyes. The ZnO-Ca nanocomposites were employed for the degradation of methyl orange dye. The reported removal efficiency of nanocomposite was 82% (Oliveira et al., 2019). A composite of ZnO was synthesized by the SnS metal sulfide. The novel preparation of ZnO-SnS nanocomposite was an effective approach to enhance photocatalytic degradation of textile dyes. The synthesized nanocomposite was employed for the degradation of rhodamine B dye. The reported removal efficiency was 95% (Wang et al., 2017). In a reported method, biosynthesis of nanocomposite was given with metal oxide. In this method, nanocomposite of chitosan-ZnO was synthesized for increasing the adsorbent properties of ZnO. The chitosan-ZnO nanocomposite was applied for the degradation of reactive black HN and reactive magenta HB dyes. The reported removal efficiency for reactive black HN and reactive magenta HB dyes was 95–99% (Abul et al., 2015). Researchers developed a new nanocomposite by using ZnO nanoparticles. Ru was employed with ZnO to synthesized composite of ZnO-Ru. The prepared nanocomposite had high rate of photocatalytic degradation of textile dyes, i.e., the reported removal efficiency was 87% for methyl orange dye (Vaiano & Iervolino, 2019). On the basis of above-reported results, it can be concluded that chitosan-ZnO nanocomposite possesses excellent degradation of dyes.

CeO_2 is another metal oxide used in the elimination of dyes from wastewater. The nanoparticles of CuO were employed with CeO_2 to synthesize a nanocomposite. The nanocomposite yields 91% removal efficiency of methyl blue (Raees et al., 2021). In another approach, nanoparticles of Y_2O_3 were used with CeO_2 for the formation of nanocomposite. The nanocomposite Y_2O_3-CeO_2 was applied for the eradication of rhodamine B textile dyes. The composite had a removal efficiency of 98% (Magdalane et al., 2017). Another study was reported in which La_2O_3 was employed with CeO_2 to yield nanocomposite. The nanocomposite was synthesized to analyze its role in the degradation of dyes. The La_2O_3-CeO_2 nanocomposite was utilized in the degradation of rhodamine B dye with the reported removal efficiency of 98% (Magdalane et al., 2018). NiO nanoparticles are also used in the treatment of wastewater. The NiO nanoparticles are doped with polyaniline in the synthesis of nanocomposite. The nanocomposite has its application in the degradation methylene blue and rhodamine B dyes. The end results showed that degradation efficiency of dyes were 94% (Taymaz et al., 2021). Cuo nanoparticles were doped with different metal oxides to enhance the degradation of textile dyes. WO_3 is a metal oxide that is applied for the removal of dyes by adsorption and photocatalytic process. WO_3 is doped with CuO nanoparticles to form a composite that enhances the photocatalytic and adsorption properties of metal oxide. The results suggested that synthesized nanocomposite degrades methyl blue dye with 90.1% removal efficiency (Dursun et al., 2020). In a method, G/CuO/Co_3O_4 nanocomposite synthesis was reported for the degradation of methyl orange and Congo red dyes. The reported results suggested that 99.85% and 99.50% degradation of dyes took place within 15 min (Mahalingam et al., 2018) (Table 11.2).

UV/visible spectroscopy has a wide range of application. Here in this paper, we studied UV/visible application in the degradation of textile dyes from industrial wastewater. In a reported method, Co metal was employed for the degradation of Murexide dye. The dye was completely degraded at UV/visible wavelength of 474–508 nm (Adekunle et al., 2020). TiO_2 is a metal oxide that is also used in the elimination of different textile dyes. TiO_2 is utilized for the degradation of methylene blue, methyl violet, and acid violet under UV/visible wavelength of 665 nm, 590 nm, and 540 nm, respectively (Rajagopal et al., 2020). In another method, the same metal oxide, i.e., TiO_2, was applied for the degradation of malachite green (MG) and methylene blue (MB) dyes. The elimination from wastewater took place at UV/visible wavelength of 528 nm (Bansal et al., 2020). Another researcher reported the degradation of AB 113 by using TiO_2 metal oxide. The AB 113 was eliminated from industrial water under UV/visible wavelength of less than 190 nm (Mortazavian et al., 2019). CoO is a metal oxide that is used in the elimination of dyes from wastewater. In a reported approach, CoO was used for the elimination of RGY dye. The dye was degraded at UV/visible wavelength of 410 nm (Cruz et al., 2019). WO_3 is a metal oxide that is also used in the degradation of textile dyes from wastewater. A researcher reported a method in which elimination of Orange II and 2-nitrophenol was reported. These dyes were degraded under the UV/visible wavelength of 480 nm (Umukoro et al., 2017).

11.6 Environmental Factor Influence

Different factors influence the degradation of textile dyes. The most important factors on which degradation mainly depends are effect of the dye concentration, pH, time, temperature, and photocatalytic activity. Herein, we look at the effect of pH on removal efficiency. In a reported method, N-doped ZnO nanoparticles were employed for the eradication of methylene blue dye. The elimination efficiency was analyzed at pH range of 3.0 to 11.0. The removal efficiencies reported at pH 3, pH 5, pH 9, and pH 11 were 80%, 85%, 90%, and 99%, respectively. Hence, it can be easily estimated that alkaline pH supports high removal efficiency (Prabakaran &

Table 11.2 Degradation of dyes under different UV/visible wavelength with removal efficiency regarding time by using metal oxide

Sr. no.	Metal oxides	Wastewater yes	Wavelength (nm)	Time	Removal efficiency	References
1	ZnO Cu@ZnO	Navy blue	599	140 min	87% 92%	Kumaran et al. (2021)
2	CeO2/CuO	Methyl blue	670	3 h	91%	Raees et al. (2021)
3	Ag_2O/ $AgO\text{-}TiO_2$	Reactive blue 220 (RB 220)	609	90 min	96%	Kulal and Kodialbail (2021)
4	Zn/Tu-TiO_2	Azo dyes	25–700	4 h	87–99%	Foulady-Dehaghi and Behpour (2020)
5	CuO/WO_3	Methylene blue	330	2 h	90.1%	Dursun et al. (2020)
6	ZnO/Ru	Methyl orange	365	4 h	87%	Vaiano and Iervolino (2019)
7	TiO/CdSe	Methylene blue	–	5 h	98.5%	Xu et al. (2019)
8	Gd/ZnO	Methylene blue	664	90 min	93%	Selvaraj et al. (2019)
9	Ca/ZnO	Methyl orange	550–660	4 h	82%	Oliveira et al. (2019)
10	TiO_2-rGO-Fe_2O_3	Methylene blue	664	6 min	99%	Banerjee et al. (2018)
11	CeO_2/La_2O_3	Rhodamine B	554	4 h	98%	Magdalane et al. (2018)
12	$G/CuO/Co_3O_4$	MO CR	–	15 min	99.85% 99.50%	Mahalingam et al. (2018)
13	ZnO-SnS	Rhodamine	420	3 h	95%	Wang et al. (2017)
14	S-TiO2	Methyl orange	430	6 h	92.8%	Yu et al. (2017)
15	CeO_2/Y_2O_3	Rhodamine B	553	180–210 min	98%	Magdalane et al. (2017)
16	$Fe_3O_4@TiO_2$/ SiO aerogel	Rhodamine B	554	60 min	97%	Li et al. (2016)
17	Nd-TiO_2-GO	Indigo carmine	610	3 h	92%	Oppong et al. (2016)
18	P_{25}/TiO_2	Rhodamine B	–	150 min	99%	Qin et al. (2009)
19	Chitosan-ZnO	Reactive black HN, reactive magenta HB	–	60 min	95–99%	Abul et al. (2015)

Pillay, 2019). In another reported method, Mn-doped ZnO metal oxide was used for the elimination of methylene blue dye. The pH effect was analyzed at a range of pH 50 to 12.0. The reported results suggested that at acidic pH removal efficiency was not high as compared to alkaline PH. The reported results suggested that maximum removal efficiency took place at pH 10 (Chanu et al., 2019). In a method, a researcher reported the degradation of methylene blue dye by using chitosan-La^{3+}-graphite composite. The pH effect was studied on the elimination efficiency. The pH range was taken in the range 3.0–11.0. It was analyzed that the rate of methylene blue degradation was high at basic pH (Sirajudheen & Meenakshi, 2019). It has been reported that at alkaline pH nanoparticles of metal oxide indicate high elimination of methylene blue dye. In another reported method (Magdalane et al., 2018), CeO_2 was employed for the degradation of organic dye rhodamine B, and the pH effect was analyzed on the removal efficiency. The reported results suggested that removal efficiency increases with increase in pH. This is due to the reason that at lower pH adsorption is poor (Magdalane et al., 2018). In another method, CeO_2-Y_2O_3 was applied for the eradication of rhodamine B dye. The pH effect was studied to analyze the better removal efficiency. The reported results showed that in acidic medium degradation efficiency was poor. It is due to decreased in the number of active sites. On other hand, degradation efficiency in alkaline pH was greatly increased (Magdalane et al., 2017).

A method was reported in which researchers used CuO metal oxides for the degradation of reactive black 5 dye. The pH effect was studied on removal efficiency. It was reported that removal efficiency was high in acidic medium, i.e., 98% at pH 4. On other hand, in alkaline medium at pH 10, the degradation efficiency was 55% (Rao et al., 2020). In another method, Fe (III) metal was applied for the removal of reactive black 5 dye, and the pH effect was studied on the removal efficiency. It was analyzed that removal efficiency was high at acidic pH. The pH increased as pH increased from 2.3 to 2.6 (Sadhu et al., 2020).

11.7 Application of Metal Oxide Nanocomposite in Dye Degradation

11.7.1 Application of ZnO Metal Oxide Nanocomposite

ZnO is the most widely used metal oxide in the elimination of toxic dyes from wastewater. ZnO is also doped with other nanoparticles to synthesize nanocomposite that gives more efficient result. Here in this paper, we review how efficiency of ZnO for degradation of dyes increases as the development of technology. In a reported method in 2004, ZnO was applied as a photocatalyst for the eradication of methylene blue and eosin Y dyes. The removal efficiency of these dyes depends on different factors such as pH, concentration of dye, and intensity of lamp. The reported results showed that degradation of eosin Y dye decreased from 93% to 82%

as the light intensity decreases but for methylene blue from 58% to 74% after 2 h. The initial concentration of dye also influences its removal. In methylene blue, as the concentration of initial dye increased from 25 to 100 mg/L, the color removal efficiency has fallen from 87% to 40% in 2 h. On other hand, for eosin Y, the removal efficiency falls from 93% to 63% as concentration increases from 25 to 50 mg/L. pH was also a parameter which affects the elimination efficiency. In the case of methylene blue as the pH increased from 5.5 to 9.7, the removal efficiency has also increased from 49% to 62% in 2 h (Chakrabarti & Dutta, 2004). Another researcher reported a method of degradation of biebrich scarlet dye with ZnO. The researcher made a comparison of degradation rate by using commercial ZnO and synthesized ZnO. The reported results suggested that 80% eradication of BS took place in 4 min by synthesized ZnO. On other hand, 50% decolorization was reported by commercial ZnO. The results showed that after 10 min, synthesized ZnO degraded almost 97% dye than commercial ZnO that degraded 84%. pH is one of the main factors that influences the removal efficiency.

The pH effect was monitored on the decolorization of biebrich dye. The removal efficiency was checked out at a pH range from 3 to 11 by keeping the dye amount constant at 25 mg/L. It was seemed that maximum elimination efficiency was obtained at 10 pH. The amount of dye was also affecting the removal efficiency. The elimination efficiency of dye decreased as the quantity of dye increased. When the concentration of dye was 5 and 25 mg/L, the removal efficiency was 100% in 8 min and 16 min, respectively. When the concentration was elevated to 50 mg/L, then the degradation took place in 20 min (Kansal et al., 2010). In another method reported (Dinesh et al., 2014), ZnO nanorods and ZnO@Ag core shell were compared for the degradation of Congo red and amido black 10B dyes. The degradation took place under the wavelength of UV/visible range. The reported results showed that 50% of degradation took place in 24 min with ZnO nanorods. On other hand, by using ZnO@Ag core shell, 50% degradation took place only in 8 min (Dinesh et al., 2014). A researcher used ZnO/γ-Fe_2O_3 nanoparticles for the elimination of methylene blue and rhodamine B dyes in terms of change in absorption at UV/visible wavelength. The reported results showed that degradation efficiency for methylene blue was calculated to be 30% and for rhodamine B to be 50% at wavelength of 668 nm and 544 nm, respectively (Hitkari et al., 2017).

11.7.2 Application of TiO_2 Metal Oxide Nanocomposite

TiO_2 is also the most significant metal oxide for the elimination of textile dyes from wastewater. A researcher (Zielińska et al., 2003) utilized TiO_2-Tytanpol A11 and TiO_2-Degussa P_{25} nanocomposite for the degradation of reactive red 198 (RR198), acid black 1 (AB1), and acid blue 7 (AB7). The reported results suggested that after 20 min of UV/visible irradiation,100% AB1 and 90% AB7 was removed (Zielińska et al., 2003). In another reported method, TiO_2 with C, N, and S tri-doped nanocomposite was synthesized to elevate the photocatalytic degradation of TiO_2. The tri-doped/TiO_2 was applied for the degradation of methylene blue dye. The reported results suggested that the efficiency of tri-doped/TiO_2 toward removal of methylene blue was 98% in 3 h. The concentration of dye played an important role toward its degradation. The photodegradation was analyzed from a concentration of 10–100 mg/L. It was noted that as concentration increased elimination efficiency decreased. The pH is also the main factor that determines the degradation efficiency. The degradation of methylene blue was determined at pH range from 3 to 11. The reported results suggested that 100% degradation was achieved at pH value 9 (Shi et al., 2011). In a published method, TiO_2 was used for the elimination of methyl blue and Congo red dyes. The reported results suggested that degradation of methyl blue took place within 420 and 700 min with efficiency of 91%. On other hand, degradation of Congo red dye takes place within 660 and 780 min with 90% (Mukhlish et al., 2013). In another reported method, Au was doped with TiO_2 to elevate the photocatalytic activity of TiO_2 for elimination of textile dyes from wastewater. Au-TiO_2 was applied for the degradation of malachite green (MG) and methylene blue (MB) dyes. The reported results showed that after time interval of 30 min, removal efficiency for MB is 59% and for MG is 26% (Bansal et al., 2020).

11.7.3 Application of CuO Metal Oxide Nanocomposite

CuO is a metal oxide that also has significant applications in industrial field. The role of CuO toward the degradation of textile dyes is also appreciable. In another method (Riaz et al., 2010), Cu-Ni/TiO_2-doped binary metal nanocomposite was synthesized for the degradation of Orange II dye at different pH. The reported results suggested that maximum degradation took place at pH 6.8. The removal efficiency at this pH was 100%. The pH is a main factor in the degradation of textile dyes. The pH results were carried out at pH range of 3–12. The maximum degradation of Orange II took place at 6.8 (Riaz et al., 2010). A researcher (Sonia et al., 2015) reported a method in which Zn was doped with CuO nanoflowers to synthesize a nanocomposite. The synthesized nanocomposite was used in the degradation of methylene blue and methylene violet dye. The reported results suggested that complete degradation of MB and MV took place in 3 h. The effect of catalyst also plays a vital role in the degradation of textile dyes. The reported results suggested that efficiency of elimination of dyes increased with the increasing concentration of doped catalyst (Sonia et al., 2015). In a reported method, CuO nanoparticles were applied for the degradation of reactive green 19A (RG19A) textile dye. The reported results suggested that removal efficiency of CuO nanoparticles was 98% within time interval of 12 h (Saratale et al., 2018). A method was reported in which the CuO was doped with different other metal oxides for the enhancement of photocatalytic degradation of Congo red dye. The CuO was doped with ZnO@CuO and Fe_3O_4@CuO. The reported results showed that removal efficiency of Congo dye with CuO, ZnO@CuO, and Fe_3O_4@CuO was 75%, 90%, and 94%, respectively (Malwal & Gopinath, 2016).

11.7.4 Application of SnO Metal Oxide Nanocomposite

SnO_2 is a metal oxide that also has an important role in the degradation of textile dyes. In a reported method, SnO_2 was used for the elimination of methylene blue from wastewater. The reported results showed that after time interval of 7.5 h 34% degradation of dye took place by the process of photolysis. Thermal treatment also affects the photocatalytic properties significantly. In another method, a researcher reported a method for the preparation of SnO_2-MgO nanocomposite by sol-gel method. The synthesized nanocomposite was applied for the degradation of methylene blue. The reported results showed that complete degradation of methylene blue took place in 40 min under sunlight. The effect of pH on photocatalytic activity of material was analyzed at different pH range from 4 to 11. The reported results showed that maximum photocatalytic degradation took place at higher pH, e.g., 8 and 9, compared to low pH 3 and 4 (Bayal & Jeevanandam, 2013). In another reported method, nanocomposite of SnO_2 was prepared. The synthesized nanoparticles were used for the degradation of methylene blue dye. The reported results suggested that removal of methylene blue took place in 4 h at wavelength of 650 nm (Bhattacharjee & Ahmaruzzaman, 2015). In another reported method, a researcher prepared SnO_2 nanoparticles by piper beetle leaf extract. The synthesized nanoparticles were used for the degradation of reactive yellow (RY-186) dye. The end results about the elimination of dye from wastewater were with 92.17% efficiency. The effect of pH and zero-point charge was also analyzed in that work. It was showed that zero zeta-point value of the nanoparticles was measured at pH 6.5. It showed that acidic pH produced substantial aggregation of nanoparticles (Singh et al., 2018).

11.7.5 Application of CeO_2 Metal Oxide Nanocomposite

CeO_2 is a metal oxide that has its application in the removal of textile dyes from wastewater. The CeO_2 nanoparticles employ with different catalyst and other metal oxides to form a nanocomposite. The nanocomposite has greater efficiency properties. A researcher reported a method in which $SrTiO_3/CeO_2$ composite was synthesized. The synthesized nanocomposite was used for the eradication of Direct Red 23 (DR-23) dye. The reported results showed that complete removal of dye took place within 60 min. The effect of pH was also a main factor on removal efficiency. The pH was selected in the range of 2.0–12.0. It was reported that worst result was shown at pH 6.0 as compared with best result shown at 12.0. It was concluded that elimination efficiency proceeded faster at alkaline pH than acidic (Song et al., 2008). In another reported method, CeO_2-MoO_3 composite was synthesized by sol-gel process applying electrospinning technique. The CeO_2-MoO_3 composite was used for the elimination of rhodamine B and safranin-T dye from industrial effluents. The reported results showed that removal efficiency of CeO_2-MoO_3 composite was 98.3% and 98.5%, respectively. The pH effect was also studied on the removal efficiency. The pH range was taken from 2.0 to 10.0. The results showed that degradation performance of ST increased as pH increased and reverse was noticed for rhodamine B (Zhao et al., 2010). In another reported method, a researcher fabricated CeO_2 by using precipitation method. This metal oxide was used for the degradation of acid orange-10 (AO-10) and Congo red (CR) dyes. The reported results suggested that CeO_2 degraded acid orange-10 (AO-10) and Congo red (CR) dyes with significant efficiency. The efficiency rate of dye elimination was 96.82% and 93.55% in time interval 60 min and 90 min, respectively (Zuas et al., 2014). In another method, a researcher focused on CeO_2/CuS nanocomposite. The synthesized nanocomposite was applied for the elimination of methylene blue (MB) dye from wastewater. The reported results showed that removal efficiency of CeO_2/CuS nanocomposite was 95% (Sabzehmeidani et al., 2019).

11.7.6 Application of NiO Metal Oxide Nanocomposite

NiO is a metal oxide that plays a vital role in the eradication of toxic dyes from industrial effluents. Different methods have been reported in which NiO metal oxide was employed for the treatment of industrial effluents. In another reported method, NiO-ZnO nanocomposite was synthesized to increase photocatalytic activity of NiO for the elimination of methylene blue. The degradation took place under UV/visible wavelength at 665 nm. The reported time for the removal of methylene blue was 80 min (Abdul Rahman et al., 2014). In another method, a researcher reported the synthesis of Ni-TiO_2 nanocomposite by hydrothermal process. The Ni-TiO_2 nanocomposite was used for the degradation of methylene blue dye. The degradation took place at 550 nm under UV/visible wavelength. The reported removal efficiency was 63% (Nakhate et al., 2010). A method was reported in which NiO nanoparticles were employed for the removal of Congo red dye from wastewater. The NiO nanocomposite was synthesized by the coprecipitation method. The NiO degrades Congo red dye in the wavelength of UV/visible light. The time interval for degradation was 160 min. The removal efficiency reported by researcher was 85% (Bhat et al., 2020).

11.7.7 Application of WO_3 Metal Oxide Nanocomposite

In recent years, WO_3 is known as the most promisable metal oxide for the degradation of dyes from wastewater. A lot of researcher worked on WO_3 to explore its role in the elimination of dyes from industrial effluents. A work was published in the first decade of twenty-first century in which synthesis of WO_3 was reported by the precipitation method. The synthesized WO_3 was used for the degradation of alizarin yellow GG dye. The degradation took place in UV/visible region at 335 nm wavelength. The reported results showed that maximum degradation took place in

20 min. The reported removal efficiency was 85% (Gondal et al., 2011). A researcher (Hunge et al., 2016) reported a method in which WO_3 thin films were prepared by pyrolysis. The thin films of WO_3 were used for the degradation of methyl blue dye. The degradation of methyl blue dye took place in UV region at 660 nm. The reported time interval was 120 min (Hunge et al., 2016). In a reported method, a researcher synthesized TiO_2-WO_3 nanocomposite by sol-gel method. The nanocomposite was used for the degradation of azo dyes, i.e., Congo red and methyl red. The degradation of azo dyes took place only in 120 min. The reported removal efficiency was 95% (Patil et al., 2019).

11.8 Conclusion

This review paper sums up the information on the elimination of textile dyes from the industrial wastewater using different metal oxides and metal oxide nanocomposite under different UV/visible wavelengths. In recent years, nanocomposites are widely used for the removal of dyes. Due to high surface area, more active sites and high cryptanalytic nanocomposites perform exceptionally well in the degradation of textile dyes. Based on literature review that is summed up in this paper, it is illustrated that nanocomposites of different metal oxide perform exceptionally well in the degradation of textile dyes. The effect of pH was also studied in the degradation of textile dyes, and it is concluded that some dyes eliminate well in acidic medium and some are in alkaline pH. In the future, green synthesis of nanocomposites will be preferred for wastewater treatment from textile dyes so no secondary pollutants will be generated.

Acknowledgments Higher Education Commission of Pakistan NRPU projects 7958 and 7964. Thanks are due to Pakistan Science Foundation Project PSF/Res/CP/C-CUI/Envr (151) for providing the funding. Furthermore, thanks are due to Pakistan Academy of Sciences Project 3-9/PAS/98 for funding.

Author Contribution Jabran Ahmad wrote paper, Muhammad Zaffar Hashmi supervised, developed idea, wrote paper, Aamer Saeed, Siwatt Pongpiachan, Xiaomei Su, Arianit A. Reka helped in revision, Zulkfil Ahmed wrote some part of paper.

Funding Higher Education Commission of Pakistan NRPU projects 7958 and 7964. Thanks to the Pakistan Science Foundation project PSF/Res/CP/C-CUI/Envr (151) for providing the funding. Furthermore, thanks are due to the Pakistan Academy of Sciences project 3-9/PAS/98 for funding.

Availability of Data and Materials This is a review paper and based on the previous published papers. Table data was generated based on the papers cited and used in the review papers. Figures were redrawn and properly acknowledged.

Ethical Approval Not applicable.

Consent to Participate Not applicable.

Consent to Publish Not applicable.

Competing of Interest None.

References

Abdul Rahman, I., Ayob, M., & Radiman S. (2014). Enhanced photocatalytic performance of NiO-decorated ZnO nanowhiskers for methylene blue degradation.

Abul, A., Samad, S., Huq, D., Moniruzzaman, M., & Masum, S. M. (2015). Textile dye removal from wastewater effluents using chitosan-ZnO nanocomposite. *Journal of Textile Science & Engineering, 5*, 2.

Acikgoz, C., Hempenius, M. A., Huskens, J., & Vancso, G. J. (2011). Polymers in conventional and alternative lithography for the fabrication of nanostructures. *Europoen Polymer Journal, 47*, 2033–2052.

Adekunle, A. S., Oyekunle, J. A., Durosinmi, L. M., Oluwafemi, O. S., Olayanju, D. S., Akinola, A. S., Obisesan, O. R., Akinyele, O. F., & Ajayeoba, T. A. (2020). Potential of cobalt and cobalt oxide nanoparticles as nanocatalyst towards dyes degradation in wastewater. *Nano-Structures & Nano-Objects, 21*, 100405.

Agustina, T. E. (2020). Study of synthetic dye removal using fenton/TiO2, Fenton/UV, and Fenton/TiO2/UV methods and the application to jumputan fabric wastewater (similarity).

Alshabanat, M. N., & AL-Anazy, M. M. (2018). An experimental study of photocatalytic degradation of congo red using polymer nanocomposite films. *2018*.

Anku, W. W., Agorku, E. S., Oppong, S. O. B., & Karikari, A. Y. (2020). MWCNTs attached neodymium doped-ZnO photocatalysts for efficient removal of dyes from wastewater. *SN Applied Sciences, 2*, 1–13.

Appavu, B., Thiripuranthagan, S., Ranganathan, S., Erusappan, E., & Kannan, K. (2018). BiVO4/N-rGO nano composites as highly efficient visible active photocatalyst for the degradation of dyes and antibiotics in eco system. *Ecotoxicology and Environmental Safety, 151*, 118–126.

Arcanjo, G. S., Mounteer, A. H., Bellato, C. R., da Silva, L. M. M., Dias, S. H. B., & da Silva, P. R. (2018). Heterogeneous photocatalysis using TiO2 modified with hydrotalcite and iron oxide under UV–visible irradiation for color and toxicity reduction in secondary textile mill effluent. *Journal of Environmental Management, 211*, 154–163.

Banerjee, S., Benjwal, P., Singh, M., & Kar, K. K. (2018). Graphene oxide (rGO)-metal oxide (TiO2/Fe3O4) based nanocomposites for the removal of methylene blue. *Applied Surface Science, 439*, 560–568.

Bansal, J., Hafiz, A., & Sharma, S. N. (2020). Photoreduction of dye with noble metal gold permeated with metal oxide titania. *Journal of Nanoscience and Nanotechnology, 20*, 3896–3901.

Bayal, N., & Jeevanandam, P. (2013). Sol–gel synthesis of SnO2–MgO nanoparticles and their photocatalytic activity towards methylene blue degradation. *Materials Research Bulletin, 48*, 3790–3799.

Bhat, S. A., Zafar, F., Mondal, A. H., Kareem, A., Mirza, A. U., Khan, S., Mohammad, A., Haq, Q. M. R., & Nishat, N. (2020). Photocatalytic degradation of carcinogenic Congo red dye in aqueous solution, antioxidant activity and bactericidal effect of NiO nanoparticles. *Journal of the Iranian Chemical Society, 17*, 215–227.

Bhattacharjee, A., & Ahmaruzzaman, M. (2015). A green and novel approach for the synthesis of SnO2 nanoparticles and its exploitation as a catalyst in the degradation of methylene blue under solar radiation. *Materials Letters, 145*, 74–78.

Bumbrah, G. S., & Sharma, R. M. (2016). Raman spectroscopy–Basic principle, instrumentation and selected applications for the characterization of drugs of abuse. *Egyptian Journal of Forensic Sciences, 6*, 209–215.

Cai, Z., Sun, Y., Liu, W., Pan, F., Sun, P., & Fu, J. (2017). An overview of nanomaterials applied for removing dyes from wastewater. *Enivironmental Science and Pollution and Research International, 24*, 15882–15904.

Chakrabarti, S., & Dutta, B. K. (2004). Photocatalytic degradation of model textile dyes in wastewater using ZnO as semiconductor catalyst. *Journal of Hazardous Materials, 112*, 269–278.

Chanu, L. A., Singh, W. J., Singh, K. J., & Devi, K. N. (2019). Effect of operational parameters on the photocatalytic degradation of Methylene blue dye solution using manganese doped ZnO nanoparticles. *Results in Physics, 12*, 1230–1237.

Cruz, J. C., Nascimento, M. A., Amaral, H. A., Lima, D. S., Teixeira, A. P. C., & Lopes, R. P. (2019). Synthesis and characterization of cobalt nanoparticles for application in the removal of textile dye. *Journal of Environmental Management, 242*, 220–228.

Dinesh, V., Biji, P., Ashok, A., Dhara, S., Kamruddin, M., Tyagi, A. K., & Raj, B. (2014). Plasmon-mediated, highly enhanced photocatalytic degradation of industrial textile dyes using hybrid ZnO@ Ag core–shell nanorods. *RSC Advances, 4*, 58930–58940.

Dos Santos, A. J., Batista, L. M. B., Martínez-Huitle, C. A., Alves, A., & Garcia-Segura, S. (2019). Niobium oxide catalysts as emerging material for textile wastewater reuse: Photocatalytic decolorization of azo dyes. *Catalysts, 9*, 1070.

Dursun, S., Koyuncu, S. N., Kaya, İ. C., Kaya, G. G., Kalem, V., & Akyildiz, H. (2020). Production of CuO–WO3 hybrids and their dye removal capacity/performance from wastewater by adsorption/photocatalysis. *Journal of Water Process Engineering, 36*, 101390.

Dutta, A. K., Ghorai, U. K., Chattopadhyay, K. K., & Banerjee, D. (2018). Removal of textile dyes by carbon nanotubes: A comparison between adsorption and UV assisted photocatalysis. *Physica E: Low-dimensional Systems and Nanostructures, 99*, 6–15.

Foulady-Dehaghi, R., & Behpour, M. (2020). Visible and solar photodegradation of textile wastewater by multiple doped TiO2/Zn nanostructured thin films in fixed bed photoreactor mode. *Inorganic Chemistry Communications, 117*, 107946.

Gnanasekaran, L., Hemamalini, R., Saravanan, R., Ravichandran, K., Gracia, F., Agarwal, S., & Gupta, V. K. (2017). Synthesis and characterization of metal oxides (CeO2, CuO, NiO, Mn3O4, SnO2 and ZnO) nanoparticles as photo catalysts for degradation of textile dyes. *Journal of Photochemistry and Photobiology B Biology, 173*, 43–49.

Gondal, M., Hayat, K., Khaled, M. M., Yamani, Z., & Ahmed, S. (2011). Photocatalytic removal of hazardous dye from water using nanostructured WO3. *International Journal Of Nanoparticles, 4*, 53–63.

Gupta, V. K., Saravanan, R., Agarwal, S., Gracia, F., Khan, M. M., Qin, J., & Mangalaraja, R. V. (2017). Degradation of azo dyes under different wavelengths of UV light with chitosan-SnO2 nanocomposites. *Journal of Molecular Liquids, 232*, 423–430.

Hasanpour, M., & Hatami, M. (2020). Photocatalytic performance of aerogels for organic dyes removal from wastewaters: Review study. *Journal of Molecular Liquids, 309*, 113094.

Hitkari, G., Singh, S., & Pandey, G. (2017). Synthesis, characterization and visible light degradation of organic dye by chemically synthesized ZnO/ γ-Fe3O4 nanocomposites. *International Journal of Advanced Research in Science, Engineering and Technology, 4*.

Hitkari, G., Sandhya, S., Gajanan, P., Shrivash, M., & Deepak, K. (2018). Synthesis of chromium doped cobalt oxide (Cr: Co3O4) nanoparticles by co-precipitation method and enhanced photocatalytic properties in the visible region. *Journal of Material Science & Engineering, 7*, 2169–2222.

Hunge, Y., Mahadik, M., Mohite, V., Kumbhar, S., Deshpande, N., Rajpure, K., Moholkar, A., Patil, P., & Bhosale, C. (2016). Photoelectrocatalytic degradation of methyl blue using sprayed WO3 thin films. *Journal of Materials Science: Materials in Electronics, 27*, 1629–1635.

Iravani, S., Korbekandi, H., Mirmohammadi, S. V., & Zolfaghari, B. (2014). Synthesis of silver nanoparticles: Chemical, physical and biological methods. *Research in Pharmaceutical Sciences, 9*, 385.

Jawaid, M., Ahmad, A., & Lokhat, D. (2019). *Graphene-based nanotechnologies for energy and environmental applications*. Elsevier.

Kansal, S. K., Ali, A. H., & Kapoor, S. (2010). Photocatalytic decolorization of biebrich scarlet dye in aqueous phase using different nanophotocatalysts. *Desalination, 259*, 147–155.

Katheresan, V., Kansedo, J., & Yon, J. L. S. Y. (2018). Efficiency of various recent wastewater dye removal methods: A review. *Journal of Environmental Chemical Engineering, 6*, 4676–4697.

Kaviyarasu, K., Magdalane, C. M., Jayakumar, D., Samson, Y., Bashir, A., Maaza, M., Letsholathebe, D., Mahmoud, A. H., & Kennedy, J. (2020). High performance of pyrochlore like Sm2Ti2O7 heterojunction photocatalyst for efficient degradation of rhodamine-B dye with waste water under visible light irradiation. *Journal of King Saud University Science, 32*, 1516–1522.

Khan, I., Saeed, K., & Khan, I. (2019). Nanoparticles: Properties, applications and toxicities. *Arabian Journal of Chemistry, 12*, 908–931.

Kulal, D., & Kodialbail, V. S. (2021). Visible light mediated photocatalytic dye degradation using Ag2O/AgO-TiO2 nanocomposite synthesized by extracellular bacterial mediated synthesis-An eco-friendly approach for pollution abatement. *Journal of Environmental Chemical Engineering, 9*, 105389.

Kumaran, V., Reshmawhahini, V., & Gomathipriya, P. (2021). Effect of Cu dopant on ZnO photocatalyst in the degradation of navy blue textile dye from synthetic wastewater. *Chemical Engineering Technology, 44*, 942–947.

Lengke, M. F., Fleet, M. E., & Southam, G. J. L. (2006). Synthesis of platinum nanoparticles by reaction of filamentous cyanobacteria with platinum (IV)– chloride complex. *Langmuir, 22*, 7318–7323.

Li, Z.-D., Wang, H.-L., Wei, X.-N., Liu, X.-Y., Yang, Y.-F., & Jiang, W.-F. (2016). Preparation and photocatalytic performance of magnetic Fe3O4@ TiO2 core–shell microspheres supported by silica aerogels from industrial fly ash. *Journal of Alloys and Compounds, 659*, 240–247.

Luza, L., Gual, A., Eberhardt, D., Teixeira, S. R., Chiaro, S. S., & Dupont, J. (2013). "Imprinting" catalytically active Pd nanoparticles onto ionic-liquid-modified Al2O3 supports. *ChemCatChem, 5*, 2471–2478.

Magdalane, C. M., Kaviyarasu, K., Vijaya, J. J., Siddhardha, B., Jeyaraj, B., Kennedy, J., & Maaza, M. (2017). Evaluation on the heterostructured CeO2/Y2O3 binary metal oxide nanocomposites for UV/Vis light induced photocatalytic degradation of Rhodamine-B dye for textile engineering application. *Journal of Alloys and Compounds, 727*, 1324–1337.

Magdalane, C. M., Kaviyarasu, K., Matinise, N., Mayedwa, N., Mongwaketsi, N., Letsholathebe, D., Mola, G., AbdullahAl-Dhabi, N., Arasu, M. V., & Henini, M. (2018). Evaluation on La2O3 garlanded ceria heterostructured binary metal oxide nanoplates for UV/visible light induced removal of organic dye from urban wastewater. *South African Journal of Chemical Engineering, 26*, 49–60.

Mahalingam, S., Ramasamy, J., & Ahn, Y.-H. (2018). Enhanced photocatalytic degradation of synthetic dyes and industrial dye wastewater by hydrothermally synthesized G–CuO–Co 3 O 4 hybrid nanocomposites under visible light irradiation. *Journal of Cluster Science, 29*, 235–250.

Malwal, D., & Gopinath, P. (2016). Enhanced photocatalytic activity of hierarchical three dimensional metal oxide@ CuO nanostructures towards the degradation of Congo red dye under solar radiation. *Catalysis Science & Technology, 6*, 4458–4472.

Martins, R. M., Rosa, A. P., & de Sousa, R. C. Adsorption of dye from wastewater effluents from textile industries using alternative adsorbents: chitosan and sawdust

Mohanraj, V., & Chen, Y. (2006). Nanoparticles-A review. *Tropical Journal of Pharmaceutical Research, 5*, 561–573.

Mortazavian, S., Saber, A., & James, D. E. (2019). Optimization of photocatalytic degradation of acid blue 113 and acid red 88 textile dyes in a UV-C/TiO2 suspension system: Application of response surface methodology (RSM). *Catalysts, 9*, 360.

Mujahid, A., & Dickert, F. L. (2012). Molecularly imprinted polymers for sensors: Comparison of optical and mass-sensitive detection. In *Molecularly imprinted sensors* (pp. 125–159). Elsevier.

Mukhlish, M. B., Najnin, F., Rahman, M. M., & Uddin, M. J. (2013). Photocatalytic degradation of different dyes using TiO2 with high surface area: A kinetic study. *Journal of Scientific Research, 5*, 301–314.

Nagajyothi, P., Prabhakar Vattikuti, S., Devarayapalli, K., Yoo, K., Shim, J., & Sreekanth, T. V. M. (2020). Green synthesis: Photocatalytic degradation of textile dyes using metal and metal oxide nanoparticles-latest trends and advancements. *Critical Reviews in Environmental Science and Technology, 50*, 2617–2723.

Nakhate, G., Nikam, V., Kanade, K., Arbuj, S., Kale, B., & Baeg, J. (2010). Hydrothermally derived nanosized Ni-doped TiO2: A visible light driven photocatalyst for methylene blue degradation. *Materials Chemistry and Physics, 124*, 976–981.

Oliveira, A. G., de Lara Andrade, J., Montanha, M. C., Lima, S. M., da Cunha Andrade, L. H., Hechenleitner, A. A. W., Pineda, E. A. G., & de DMF, O. (2019). Decontamination and disinfection of wastewater by photocatalysis under UV/visible light using nano-catalysts based on ca-doped ZnO. *Journal of Environmental Management, 240*, 485–493.

Oppong, S. O.-B., Anku, W. W., Shukla, S. K., Agorku, E. S., & Govender, P. (2016). Photocatalytic degradation of indigo carmine using Nd-doped TiO 2-decorated graphene oxide nanocomposites. *Journal of Sol-Gel Science and Technology, 80*, 38–49.

Park, H.-H., Zhang, X., Choi, Y.-J., Park, H.-H., & Hill, R. H. (2011). Synthesis of Ag nanostructures by photochemical reduction using citrate-capped Pt seeds. *Journal of Nanomaterials, 2011*.

Patil, S., Deshmukh, S., More, K., Shevale, V., Mullani, S., Dhodamani, A., & Delekar, S. (2019). Sulfated TiO2/WO3 nanocomposite: An efficient photocatalyst for degradation of Congo red and methyl red dyes under visible light irradiation. *Materials Chemistry and Physics, 225*, 247–255.

Pentassuglia, S., Agostino, V., & Tommasi, T. (2018). EAB – Electroactive biofilm: A biotechnological resource.

Prabakaran, E., & Pillay, K. (2019). Synthesis of N-doped ZnO nanoparticles with cabbage morphology as a catalyst for the efficient photocatalytic degradation of methylene blue under UV and visible light. *RSC Advances, 9*, 7509–7535.

Qin, X., Jing, L., Tian, G., Qu, Y., & Feng, Y. (2009). Enhanced photocatalytic activity for degrading Rhodamine B solution of commercial Degussa P25 TiO2 and its mechanisms. *Journal of Hazardous Materials, 172*, 1168–1174.

Raees, A., Jamal, M. A., Ahmed, I., Silanpaa, M., & Saad Algarni, T. (2021). Synthesis and characterization of CeO2/CuO nanocomposites for photocatalytic degradation of methylene blue in visible light. *Coatings, 11*, 305.

Rajagopal, S., Paramasivam, B., & Muniyasamy, K. (2020). Photocatalytic removal of cationic and anionic dyes in the textile wastewater by H2O2 assisted TiO2 and micro-cellulose composites. *Separation and Purification Technology, 252*, 117444.

Rao, M. P. C., Kulandaivelu, K., Ponnusamy, V. K., Wu, J. J., & Sambandam, A. (2020). Surfactant-assisted synthesis of copper oxide nanorods for the enhanced photocatalytic degradation of reactive black 5 dye in wastewater. *Environmental Science and Pollution Research, 27*, 17438–17445.

Riaz, N., Chong, F. K., Dutta, B. K., Khan, S., & Nurlaela, E. (2010). Photocatalytic degradation of Orange II using bimetallic Cu-Ni/TiO2 photocatalysts.

Sabzehmeidani, M. M., Karimi, H., & Ghaedi, M. (2019). Visible light-induced photo-degradation of methylene blue by n–p heterojunction CeO2/CuS composite based on ribbon-like CeO2 nanofibers via electrospinning. *Polyhedron, 170*, 160–171.

Sadhu, S. P., Ruparelia, J., & Patel, U. D. (2020). Homogeneous photocatalytic degradation of azo dye reactive black 5 using Fe (III) ions under visible light. *Environmental Technology*, 1–8.

Sahoo, P. K., Kamal, S. K., Premkumar, M., Kumar, T. J., Sreedhar, B., Singh, A., Srivastava, S. K., & Sekhar, K. C. (2009). Synthesis of tungsten nanoparticles by solvothermal decomposition of tungsten hexacarbonyl. *International Journal of Refractory Metals and Hard Materials, 27*, 784–791.

Saito, M., Yasukawa, K., Umeda, T., & Aoi, Y. (2008). Copper nanoparticles fabricated by laser ablation in polysiloxane. *Optical Materials, 30*, 1201–1204.

Saratale RG, Ghodake GS, Shinde SK, Cho S-K, Saratale GD, Pugazhendhi A, Bharagava RN JJoem (2018). Photocatalytic activity of CuO/Cu (OH) 2 nanostructures in the degradation of reactive green 19A and textile effluent, phytotoxicity studies and their biogenic properties (antibacterial and anticancer). Journal of Environmental Management 223, 1086-1097.

Selvaraj, S., Mohan, M. K., Navaneethan, M., Ponnusamy, S., & Muthamizhchelvan, C. (2019). Synthesis and photocatalytic activity of Gd doped ZnO nanoparticles for enhanced degradation of methylene blue under visible light. *Materials Science in Semiconductor Processing, 103*, 104622.

Shi, Y. D., Guo, Q., & Xie, Y. S. (2011). The preparation of C, N, S tri-doped TiO2 and visible-light photo-degradation of methylene blue and dyes. *Advanced Materials Research, 287–290*, 1735–1743.

Singh, J., Kaur, N., Kaur, P., Kaur, S., Kaur, J., Kukkar, P., Kumar, V., Kukkar, D., & Rawat, M. (2018). Piper betle leaves mediated synthesis of biogenic SnO2 nanoparticles for photocatalytic degradation of reactive yellow 186 dye under direct sunlight. *Environmental Nanotechnology, Monitoring & Management, 10*, 331–338.

Sirajudheen, P., & Meenakshi, S. (2019). Facile synthesis of chitosan-La3+-graphite composite and its influence in photocatalytic degradation of methylene blue. *International Journal of Biological Macromolecules, 133*, 253–261.

Smrithi, S., Kottam, N., Arpitha, V., Narula, A., Anilkumar, G., & Subramanian, K. (2020). Tungsten oxide modified with carbon nanodots: Integrating adsorptive and photocatalytic functionalities for water remediation. *Journal of Science Advanced Materials and Devices, 5*, 73–83.

Song, S., Xu, L., He, Z., Ying, H., Chen, J., Xiao, X., & Yan, B. (2008). Photocatalytic degradation of CI direct red 23 in aqueous solutions under UV irradiation using SrTiO3/CeO2 composite as the catalyst. *Journal of Hazardous Materials, 152*, 1301–1308.

Sonia, S., Annsi, I. J., Kumar, P. S., Mangalaraj, D., Viswanathan, C., & Ponpandian, N. (2015). Hydrothermal synthesis of novel Zn doped CuO nanoflowers as an efficient photodegradation material for textile dyes. *Materials Letters, 144*, 127–130.

Taymaz, B. H., Eskizeybek, V., & Kamış, H. (2021). A novel polyaniline/NiO nanocomposite as a UV and visible-light photocatalyst for complete degradation of the model dyes and the real textile wastewater. *Environmental Science and Pollution Research, 28*, 6700–6718.

Tojo, C., de Dios, M., & Barroso, F. (2011). Surfactant effects on microemulsion-based nanoparticle synthesis. *Materials (Basel), 4*, 55–72.

Umukoro, E. H., Peleyeju, M. G., Ngila, J. C., & Arotiba, O. A. (2017). Towards wastewater treatment: Photo-assisted electrochemical degradation of 2-nitrophenol and orange II dye at a tungsten trioxide-exfoliated graphite composite electrode. *Chemical Science, 317*, 290–301.

Vaiano, V., & Iervolino, G. (2019). Photocatalytic removal of methyl orange azo dye with simultaneous hydrogen production using Ru-modified Zno photocatalyst. *Catalysts, 9*, 964.

Wang, H., & Chu, P. K. (2013). Surface characterization of biomaterials. In *Characterization of biomaterials* (pp. 105–174). Elsevier.

Wang, L., Zhai, H., Jin, G., Li, X., Dong, C., Zhang, H., Yang, B., Xie, H., & Sun, H. (2017). 3D porous ZnO–SnS p–n heterojunction for visible light driven photocatalysis. *Physical Chemistry Chemical Physics, 19*, 16576–16585.

Wu, D., Li, C., Zhang, D., Wang, L., Zhang, X., Shi, Z., & Lin, Q. (2019). Enhanced photocatalytic activity of Gd3+ doped TiO2 and Gd2O3 modified TiO2 prepared via ball milling method. *Journal of Rare Earths, 37*, 845–852.

Xu, D., Liu, B., Zou, W., Wang, H., & Zhang, C. (2019). In situ synthesis of TiO2 nanosheets@ CdSe nanocomposites and the improved photocatalytic performance on removal of methylene blue. *Applied Surface Science, 487*, 91–100.

Yadav, T. P., Yadav, R. M., & Singh, D. P. (2012). Mechanical milling: A top down approach for the synthesis of nanomaterials and nanocomposites. *Nanoscience and Nanotechnology, 2*, 22–48.

Yu, X., Wu, P., Qi, C., Shi, J., Feng, L., Li, C., & Wang, L. (2017). Ternary-component reduced graphene oxide aerogel constructed by g-C3N4/BiOBr heterojunction and graphene oxide with enhanced photocatalytic performance. *Journal of Alloys and Compounds, 729*, 162–170.

Zainal, Z., Hui, L. K., Hussein, M. Z., Taufiq-Yap, Y. H., Abdullah, A. H., & Ramli, I. (2005). Removal of dyes using immobilized titanium dioxide illuminated by fluorescent lamps. *Journal of Hazardous Materials, 125*, 113–120.

Zhao, S., Li, J., Wang, L., & Wang, X. (2010). Degradation of rhodamine B and safranin-T by MoO3: CeO2 nanofibers and air using a continuous mode. *CLEAN – Soil Air Water, 38*, 268–274.

Zielińska, B., Grzechulska, J., & Morawski, A. W. (2003). Photocatalytic decomposition of textile dyes on TiO2-Tytanpol A11 and TiO2-Degussa P25. *Journal of Photochemistry and Photobiology A Chemistry, 157*, 65–70.

Zuas, O., Abimanyu, H., & Wibowo, W. (2014). Synthesis and characterization of nanostructured CeO2 with dyes adsorption property. *Processing and Application of Ceramics, 8*, 39–46.

Chapter 12
Health Risk Assessment of Heavy Metals in Wastewater to Livestock from Punjab Province, Pakistan

Muhammad Usman Khan, Muhammad Zaffar Hashmi, Riffat Naseem Malik, and Atif Kamal

12.1 Introduction

Food crop contamination via wastewater/or contaminated soil with heavy metals results in a health hazard risk consequences for human and livestock (animals) via their consumption (Huang et al., 2014; Ondo et al., 2013; Valko et al., 2006). The food safety for livestock is an underestimated problem worldwide, especially in developing countries like Pakistan, where the dairy products (meat and milk) of the livestock are used by the human population.

Wastewater irrigation (Ww_i) is considered as an important source of heavy metal accumulation in the soil; food crops further relocate these metals into the food chain posing a severe hazards to livestock and human (Abbasi et al., 2013; Haiyan & Stuanes, 2003; Hashmi et al., 2013; Muchuweti et al., 2006). Estimates overview that wastewater is being used for the irrigation of around 20 million hectares of land throughout the world. In Pakistan, about 172 billion m^3 of water is available for irrigation which is insufficient to meet all the needs of the agriculture sector (Abbas et al., 2007). In many cities therefore, shortage of water supplies is met by the application of wastewater (Ww). Ww_i increases metal concentrations in food crops/foodstuffs which is a serious risk factor for both human and animals (Livestock) (Singh et al., 2010). Livestock reared on such wastewater-contaminated vegetables and cereal crop (*Triticum aestivum*) has accumulated parts of heavy metals in their meat and milk products. Intake of this contaminated food crops is the most direct exposure route for heavy metals to both human and livestock, i.e., through dairy

M. U. Khan · R. N. Malik (✉) · A. Kamal
Faculty of Biological Sciences, Department of Environmental Sciences, Environmental Biology and Ecotoxicology Laboratory, Quaid-i-Azam University, Islamabad, Pakistan

M. Z. Hashmi (✉)
Institute of Molecular Biology and Biotechnology, The University of Lahore, Lahore, Pakistan

H. Alshemmari et al. (eds.), *Contaminated Land and Water*, Emerging Contaminants and Associated Treatment Technologies, https://doi.org/10.1007/978-3-031-65129-8_12

product (milk and meat) consumption (Hernández-Martínez & Navarro-Blasco, 2012; Loutfy et al., 2006; Zhuang et al., 2009). Health risks associated with heavy metal exposure include mutation and adverse carcinogenic effects in livestock due to their persistent and nonbiodegradable nature (Al-Othman & Naushad, 2011; Oves et al., 2012). Even essential elements like Zn, Mn, Ni, and Cu which are commonly known as micronutrients (Singh et al., 2010) become toxic when they get accumulated in high concentrations in human food and feed of animal (Zhuang et al., 2009).

The edible parts of selected food crops (vegetables and *Triticum aestivum*) as shown in Table 12.1 are also being used on regular basis to feed domesticated livestock in Pakistan. These food crops are irrigated with Ww from most polluted streams, i.e., Nullah Aik and Nullah Palkhu, that received a huge amount of industrial, municipal, and agricultural runoff from Sialkot and Wazirabad cities (Qadir & Malik, 2011; Qadir et al., 2008). These cities are renowned worldwide for their tanneries, pharmaceuticals, surgical, sports, and cutlery industries (Ali et al., 2013). Among the studied food crops, *Triticum aestivum* are consumed in large proportion by livestock as compared to all other vegetables. This is also accounted as an indirect risk factor for human health, because there may be considerable accumulation of metals in livestock (especially in milk and meat) via selected food crops consumption that are subsequently used by human being. Livestock manure contaminated with heavy metal is used as fertilizers in agricultural lands that pose a threat to the soil and the food crops grown on such contaminated soil. This aspect represents a serious environmental issue which is needed to be addressed; otherwise, our whole food chain will be contaminated. This study aims to investigate the concentration of heavy meals in edible parts (EP_S) of food crops (vegetables and *Triticum aestivum*) irrigated with wastewater and tube well water. The main focus of this study was to highlight the potential risks to livestock of heavy metal via selected food crops (vegetables and *Triticum aestivum*) consumption.

Table 12.1 Showing edible parts of different food crops feed to livestock

Species	Abbreviation	English name	Families	Edible parts for livestock
Allium cepa L.	*A.c*	Onion	Liliaceae	Leaves
Abelmoschus esculentus L.	*A.e*	Lady's finger	Malvaceae	Leaves, young stem
Allium sativum L.	*A.s*	Garlic	Liliaceae	Leaves
Daucus carota L.	*D.c*	Carrot	Apiaceae	Leaves
Raphanus sativus L.	*R.s*	Radish	Brassicaceae	Leaves
Solanum melongena L.	*S.m*	Brinjal	Solanaceae	Leaves, stem
Lycopersicum esculentum L.	*L.e*	Tomato	Solanaceae	Leaves, stem
Vigna radiata	*V.r*		Fabaceae	Leaves, stem
Triticum aestivum L.	*T.a*	Wheat	Poaceae	Stem, leaves

12.2 Materials and Methods

12.2.1 Sampling Strategy

Sialkot and Wazirabad are renowned as the industrial cities located between 32° 24′N and 32° 37′ N latitude and 73° 59′E and 75° 20′E longitude in the east of Pakistan and 32° 27′ 0″N and 74° 7′ 0″ E of Punjab, respectively. Food crop (vegetables and *Triticum aestivum*) samples were collected in the vicinity from industrial areas of Sialkot and Wazirabad, Punjab Province, Pakistan. Sampling sites were divided into wastewater site (Ww_s) and tube well water (Tw_s) site along the bank of Nullah Aik and Nullah Palkhu streams. These are renowned as wastewater streams polluted with urban, industrial, and agricultural effluents and used for irrigation practices of agricultural land of the aforementioned cities. The selected food crops especially *Triticum aestivum* was the most popular feed of livestock that were consumed by the local livestock population of study area. Four subsampling sites were selected from Nullah Aik and Palkhu (n = 2 for each Ww_s and Tw_s) for food crops selection as shown in Fig. 12.1.

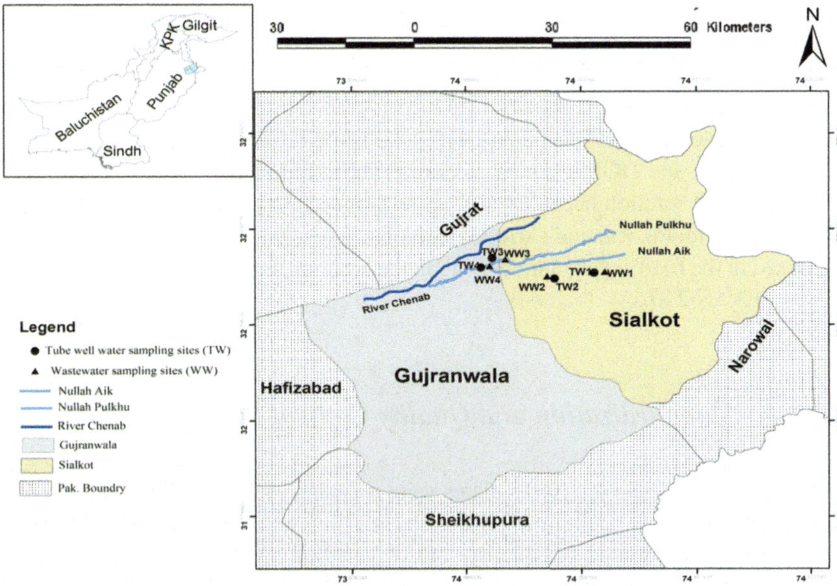

Fig. 12.1 Map of the study area showing sampling sites irrigated with wastewater and tube well water

12.2.2 Food Crops Sampling

A total of eight vegetable species, viz., *Abelmoschus esculentum* L. (*A.e*), *Allium cepa* L. (*A.c*), *Solanum melongena* L. (*S.m*), *Allium sativum* L. (*A.s*), *Daucus carota* L. (*D.c*), *Raphanus sativus* L. (*R.s*), *Lycopersicum esculentum* L. (*L.e*), *Vigna radiate* L. (*V.r*), and one cereal crop, i.e., *Triticum aestivum* L. (*T.a*), were collected from wastewater and tube well water sites. A total of 72 food crops sample (n = 36 for wastewater and n = 36 for tube well water sites) were collected randomly, i.e., one each food crop species from a total of eight sites (four sites each of wastewater and tube well water). All food crops sample were kept in polythene bags and preserved at 4 °C in a freezer. Each food crop species was collected as a composite sample of two to five subsamples from the aforementioned sites. The parts of selected food crop samples consumed by livestock as shown in Table 12.1 were preceded further for experimental analysis. The consumed parts of food crops were washed with tape and distilled water to remove all the adhered dust particles and dried overnight. Finally, the food crop samples were oven-dried at 70–80 °C for 24 h until complete dryness. Dried samples were further pulverized by micro hammer mill (Retsch, P-318, No. 611771), sieved with mesh (2 mm), and stored in small polythene zip bags at ambient temperature till further analysis.

12.2.3 Food Crops Analysis

A 0.5 g of each food crops sample was added to a flask with a triacid mixture of HNO_3, H_2SO_4, and $HCLO_4$ in a 5:1:1 ratio. The solution was further heated up to 80 °C until the solution became transparent. The selected heavy metals, i.e., Cr, Ni, Fe, Cu, Co, Mn, Pb, Zn, and Cd, and, macronutrients, i.e., K, Na, Mg, and Ca, were analyzed in the food crops sample using flame atomic absorption spectrophotometer (Varian FAAS-240).

12.2.4 Standardization and Quality Control Parameter

All the glasswares and the plasticwares were rinsed at each pre- and post-analytical stages with 10% HNO_3 solution followed by washing with deionized distilled water. For the quality assurance and quality control, certified standard and chemicals used for the preparation of samples were of analytical grade.

Standard stock solutions of 1000 mg/L were prepared for each of the selected heavy metal from certified standard of Fluka Kamica (Buchs, Switzerland). To check the precision and accuracy of the instrument, after every five samples, a blank and 2.5 and 5 μg/L concentrations of standard of the respective metals were analyzed. Each analysis included three replicates and average values which were used for further analysis. Recovery rates of >90% were observed as final for the selected metals. A blank sample was run after every five samples. For quality data assurance, each sample was analyzed in triplicate. After every five samples, three standards, one blank, and two of 2.5 and 5.0 μg/L concentrations of respective metal were analyzed for atomic absorption. The reproducibility was found to be >90% confidence level. Therefore, an average value of three replicates was used for interpretation of results. Standard solutions (1000 mg/L) of all respective metals were prepared by dilution of certified standard (NIST-SRM, 1570a for spinach leaves) of corresponding metal ions with deionized distilled water.

12.2.5 Livestock Risk Assessment Strategy or Livestock Risk via Contaminated Food Crops Consumption

The edible parts of different food crops mentioned in Table 12.1 were feed to livestock as diet. A total of 130 questionnaires of livestock including buffalo (45), cow (38), and goat (47) from Sialkot and Wazirabad cities were carried out from local inhabitant (owner) of the domestic livestock. The main information in the questionnaire was each livestock age, body weight, and daily consumption of food crops, i.e., vegetables and cereal crop (wheat), were investigated. The nonedible parts of all the vegetables and wheat straws (nonedible parts, i.e., stem, leaves, spikes) were the most popular feed to livestock in study areas and particularly in whole Pakistan. These aforementioned parts of food crops were not used by human in their diet. The daily intakes of metal in livestock from the selected food crops consumption were given in Eq. 12.1:

$$\text{DIML} = \text{Cfc} \times \text{Cf} \times \text{Dfc} / \text{Bw} \tag{12.1}$$

where Cfc represents heavy metal concentration in food crops, Cf conversion factor of fresh to dry weight of food crops, Dfc daily ingestion of food crops, and Bw body weight. This shows the higher the metal intake, the higher will be the risks to livestock from such food crops consumption and vice versa.

12.2.6 Statistical Analysis

All data was analyzed on SPSS 13.0 and Microsoft Excel. Mean heavy metals between the same species taken from two different sites were compared using independent sample t test. All the effects are reported at $P < 0.05$. Principal component analysis (PCA) was used for source identification.

12.3 Results and Discussion

In this study, the comparative among the vegetables from Tw_s and Ww_s showed significantly higher localization of heavy metals in plant species from Ww_s than Tw_s food crops species (Table 12.2).

12.3.1 Heavy Metal Concentration in the Edible Parts of Food Crops Feed to Livestock

Tables 12.3 and 12.4 summarize and evaluate metal concentration with their range, mean, and standard deviation in different food crops irrigated with tube well water and wastewater, respectively. Metal concentration in overall food crops irrigated with Tw was in the order of K>Ca>Na>Fe>Mg>Pb>Mn>Zn>Co>Ni>Cr>Cu>Cd, while pattern of metal concentration in overall food crops irrigated with Ww followed the order K>Ca>Na>Fe>Mg>Pb>Mn>Zn>Co>Ni>Cd>Cr>Cu. Overall metal concentration in food crops species grown on Tw showed the trends as D.c>A.c>V.r>S.m>T.a>R.s>A.e>A.s>L.e, while food crops grown on Ww followed the pattern of metal concentration as A.c>S.m>D.c>L.e>A.e>A.s>R.s>V.r>T.a.

Mean metal concentration in overall food crops irrigated with Ww was higher as compared to the food crops irrigated with Tw. Mean metal concentration was higher in Cr (A.e), Ni (S.m), Fe (S.m), Ca (V.r), and Mn (D.c, A.e, V.r) as compared to Ww-irrigated food crops. The maximum allowable limit (MAL) in food crops for metals, viz., Mn, Ni, Cd, Pb, Cr, Fe, Cu, and Zn, are 500, 66.9, 0.2, 0.3, 2.3, 425, 40, and 60 mg/kg, respectively (Codex Alimentarius Commission, 2001; WHO/FAO, 2007).

All the edible parts of food crops irrigated with wastewater and tube well water surpassed the MAL for Pb and Cd. Cr concentrations were above the MAL in all the food crops except *Lycopersicum esculentum*, *Raphanus sativus*, and *Daucus carota* when irrigated with wastewater. The tube well water-irrigated food crops have higher Cr concentration only in *Abelmoschus esculentum* that are attributable to the leaching of metals from wastewater because the food crops samples were collected in the vicinity of wastewater streams (Nullah Aik and Palkhu); therefore, leaching of metals from wastewater might be responsible into the underground water (tube well water) (Rattan et al., 2005). Due to such constant wastewater practices on soil, their retention capacity decreases, due to which the heavy metals leached into the ground water (Jan et al., 2010). Another possible reason for their toxicity in food crops and their potential threat to livestock and human attribute to their nonessential nature in food crops and their permissible limits as compared to other metals is very low (Saha & Zaman, 2011). All other metals except Pb, Cd, and Cr were recommended as safe in all food crops irrigated with wastewater and tube well water. High-fertilizer application especially phosphate fertilizer is the potential Cd source

Table 12.2 Comparative results of heavy metals detected in the individual species collected from wastewater and tub well water-irrigated sites

		Species heavy metals					Macronutrients							
		Cr	Ni	Fe	Cu	Co	Mn	Pb	Zn	Cd	K	Na	Mg	Ca
Allium sativum	*P*-value	NS	NS	<0.001	NS	NS	NS	NS	NS	0.001	0.03	NS	NS	0.025
Daucus carota	*P*-value	NS	0.019	0.041	NS	NS	NS	NS	NS	NS	NS	0.04	NS	NS
Raphanus sativus	*P*-value	0.005	0.001	0.001	<0.001	0.003	NS	0.001	0.009	<0.001	NS	NS	<0.001	NS
Lycopersicum esculentum	*P*-values	NS	0.001	0.001	NS	0.033	NS	0.014	0.001	0.002	0.018	0.001	0.005	0.021
Triticum aestivum	*P*-value	0.05	0.001	NS	NS	0.024	NS	0.004	NS	<0.001	NS	NS	NS	0.005
Abelmoschus esculentum	*P*-value	0.008	0.02	NS	NS	NS	NS	0.004	0.016	<0.001	NS	0.048	0.024	NS
Allium cepa	*P*-value	NS	NS	NS	NS	0.036	<0.001	0.002	0.026	<0.001	NS	NS	0.038	0.001
Solanum melongena	*P*-value	NS	NS	NS	0.013	NS	0.01	NS	NS	0.01	NS	<0.001	<0.001	<0.001
Vigna radiate	*P*-value	NS	0.008	NS	NS	NS	0.005	NS	0.002	<0.001	NS	NS	0.002	0.001

Table 12.3 Metal concentration (mg/kg) in the edible parts of food crops feed to livestock grown on tube well water

Species	statistic	Cr	Ni	Fe	Cu	Co	Mn	Pb	Zn	Cd	K	Na	Mg	Ca
A.s	Mean ± std	1.48 ± 0.78	2.9 ± 0.55	155.1 ± 16.44	0.53 ± 0.13	3.32 ± 1.27	4.74 ± 1.4	23 ± 8.3	6.15 ± 2.68	1.06 ± 0.50	259.9 ± 71.7	176.1 ± 65.9	46.60 ± 12.26	150.76 ± 30
	Min-max	0.95–2.65	2.55–3.71	135.7–174.1	0.25–0.8	0.35–6.45	2.95–6.1	15.8–32.5	3.87–9.825	0.73–1.8	187–334	92.43–252.1	36.23–64.36	108.45–179.4
D.c	Mean ± std.	0.43 ± 0.16	2.03 ± 0.4	43.34 ± 11.30	0.38 ± 0.12	2.6 ± 0.90	25.9 ± 2.1	29.0 ± 5.7	9.9 ± 2.99	0.41 ± 0.14	452 ± 163.2	227.9 ± 95.3	78.2 ± 12.6	345.71 ± 83.7
	Min-max	0.12–0.73	1.61–2.6	32.647–59.25	0.27–0.55	0.55–4.45	23.64–28.6	20.87–34	6.81–13.8	0.19–0.75	177.3–798	144.89–403.3	69.73–96.48	296.83–470.9
R.s	Mean ± std.	0.46 ± 0.14	2.99 ± 0.6	40.5 ± 8.19	0.59 ± 0.10	2.99 ± 0.7	4.95 ± 2.5	32.7 ± 4.9	8.86 ± 2.0	0.87 ± 0.19	354.3 ± 122	124.5 ± 71.2	78.8 ± 12.47	288 ± 41.37
	Min-max	0.08–0.9	2.3–3.66	33.45–52.05	0.31–0.95	0.14–8.45	2.24–8.05	25.9–37.5	6.88–11.02	0.65–1.1	201.5–470.5	24.11–178.6	60.4–87.2	228.3–316.05
L.e	Mean ± std.	0.7 ± 0.09	1.5 ± 0.2	63.5 ± 9.25	1.2 ± 0.27	3.43 ± 0.32	10.02 ± 2	25.07 ± 1.0	8.57 ± 0.8	0.66 ± 0.15	101.72 ± 7.2	126.1 ± 4.7	81.5 ± 5.88	154.36 ± 6.43
	Min-max	0.32–1.23	1.4–1.78	52.34–74.87	0.71–1.75	2.99–3.76	7.93–12.5	22.7–26.4	7.46–9.32	0.54–0.89	96.83–112.3	119.88–130.3	75.3–89.33	149.39–163.8
T.a	Mean ± std.	1.4 ± 0.7	1.8 ± 0.5	180.1 ± 27.4	1.88 ± 0.5	4.78 ± 1.64	16.3 ± 3.2	36.6 ± 2.8	11.5 ± 3.03	0.24 ± 0.08	274.75 ± 99.9	112.6 ± 4.44	54.2 ± 13.2	280.3 ± 10.1
	Min-max	0.15–3.1	1.11–2.2	149–207	1.44–2.55	2.43–6.15	12.25–19.5	34–40.5	8.3–15.615	0.05–0.39	127–365.5	105.98–15.29	38.21–66.2	271.9–294.9
A.e	Mean ± std.	3.65 ± 1.47	2.3 ± 0.76	59.83 ± 7.44	0.122 ± 0.1	4.4 ± 3.96	13.9 ± 3.99	27.2 ± 7.7	7.66 ± 1.48	0.67 ± 0.09	348.7 ± 95.2	108 ± 85.51	70.82 ± 14.49	251.5 ± 96.7
	Min-max	2.15–5.4	1.35–3.1	50.45–67.55	0.04–0.25	1.2–9.6	10.93–19.4	11–46.5	3.815–11.3	0.25–0.9	118.36–589	23.9–217.14	58.4–90.3	128.9–431.5
A.c	Mean ± std.	1.03 ± 0.29	2.76 ± 0.8	86.2 ± 45.2	0.53 ± 0.12	4.9 ± 1.2	2.4 ± 0.31	30.8 ± 4.0	7.59 ± 1.23	1.43 ± 0.27	571.2 ± 166.3	131.3 ± 55.21	126.4 ± 13.2	205.95 ± 91.5
	Min-max	0.38–1.75	2.1–3.89	52.28–148.9	0.2–0.95	2.78–9.65	2.05–2.8	24.5–35	6.56–9.365	1.13–1.75	387–791	100.6–214	108.–137.5	110.5–337.25
S.m	Mean ± std.	0.3 ± 0.08	5.7 ± 1.1	180.96 ± 11.59	0.58 ± 0.15	4.43 ± 0.90	5.04 ± 0.8	44.2 ± 7.1	12.96 ± 2.4	0.68 ± 0.19	476.6 ± 43.8	45.8 ± 6.40	37.91 ± 3.5	186.6 ± 14.6
	Min-max	0.12–0.44	5–7.23	165.69–193.6	0.38–0.74	3.38–5.57	4.16–5.94	37.28–52	9.66–15.37	0.29–1.67	437.46–533	39.36–52.95	34.56–41.5	173.16–202.7
V.r	Mean ± std.	1.06 ± 0.25	2.7 ± 0.4	94.65 ± 27.09	0.815 ± 0.1	2.56 ± 1.2	13.8 ± 2.0	26.6 ± 2.9	9.5 ± 1.03	0.26 ± 0.08	356.5 ± 107.6	166 ± 11.6	86.38 ± 7.65	297.1 ± 90.2
	Min-max	0.75–1.29	2.36–3.2	66.45–119.23	0.69–0.9	1.75–4.25	12.4–16.7	22.8–29.7	8.29–10.75	0.15–0.45	231.15–484	149.2–175.4	75.27–92.43	221.4–410.3

Abelmoschus esculentum (*A.e*), *Allium cepa* (*A.c*), *Solanum melongena* (*S.m*), *Allium sativum* (*A.s*), *Daucus carota* (*D.c*), *Raphanus sativus* (*R.s*), *Lycopersicum esculentum* (*L.e*), *Vigna radiata* (*V.r*), and *Triticum aestivum* (*T.a*)

Table 12.4 Metal concentrations (mg/kg) in the edible parts of food crops feed to livestock grown on wastewater

Species	statistic	Cr	Ni	Fe	Cu	Co	Mn	Pb	Zn	Cd	K	Na	Mg	Ca
A.s	Mean ± std.	2.72 ± 1.0	4.67 ± 1.5	286.7 ± 23.9	1.86 ± 1.1	5.6 ± 4.96	15.7 ± 2.9	41.2 ± 16.9	12.3 ± 5.8	3.89 ± 0.7	522.6 ± 107.6	235.8 ± 83.9	99.8 ± 28.4	390.95 ± 99.55
	Min-max	1.78–3.85	3.00–5.95	265.5–314.3	1.11–3.40	1.05–10.2	12.4–18.9	25.34–58	7.52–20.3	3.32–4.85	307–689.5	111.3–289.7	52.9–188.36	244.5–512.69
D.c	Mean ± std.	2.04 ± 0.8	4.66 ± 1.6	99.03 ± 69.1	1.04 ± 0.6	5.62 ± 4.3	17.02 ± 6.1	45.6 ± 4.7	15.4 ± 1.5	3.37 ± 0.50	600.9 ± 280.6	354.9 ± 126	146.8 ± 47.7	499.42 ± 134.9
	Min-max	1.05–2.9	3.50–7.0	54.35–202	0.55–1.95	1.80–9.9	10.9–24.5	39.57–50	14–16.88	2.80–3.86	206.5–809.5	232.4–497.7	115.6–217.2	345.1–662.35
R.s	Mean ± std.	2.15 ± 1.9	4.56 ± 0.6	53.4 ± 6.9	1.02 ± 0.5	4.69 ± 3.9	10.8 ± 1.16	43.4 ± 8.46	14.2 ± 1.5	3.79 ± 0.47	543.8 ± 132.1	281.3 ± 104.5	115 ± 5.2	421.7 ± 18.4
	Min-max	0.58–4.85	3.85–5.0	46.90–62.82	0.55–1.8	1.89–10.4	9.15–11.6	38.49–56	122–15.5	3.19–4.32	354.43–644.5	71.04–401.37	108.9–120.9	395.67–436.2
L.e	Mean ± std.	2.21 ± 1.2	4.60 ± 0.5	310.2 ± 76.6	1.82 ± 1.4	7.98 ± 3.3	10.18 ± 3.0	43.7 ± 10.6	15.4 ± 2.1	2.42 ± 0.22	483.2 ± 135.6	304.3 ± 63	202.4 ± 55.5	399.3 ± 156.8
	Min-max	1.15–4.0	4.10–5.23	223.1–386.9	0.15–3.50	4.96–11.3	6.05–13.15	29–51.50	13.5–18	2.15–2.65	302.32–798.5	214.9–361.7	158.–279.1	203.55–550.55
T.a	Mean ± std.	3.36 ± 0.9	4.0 ± 0.6	260.9 ± 96.9	2.39 ± 0.93	7.54 ± 0.8	21.3 ± 6.75	51.9 ± 5.96	13.9 ± 3.2	2.16 ± 0.2	357.8 ± 104.6	209.9 ± 133.7	67.8 ± 9.21	306.0 ± 6.1
	Min-max	2.10–4.3	3.21–4.4	158.3–395.5	0.80–3.15	6.50–8.40	11.85–27.8	46.0–59.5	11.2–18.3	1.95–2.4	211–444.5	127.9–407.8	55.9–78.15	300.65–312.5
A.e	Mean ± std.	3.35 ± 1.6	4.65 ± 1.2	98.74 ± 23.8	1.62 ± 0.97	6.4 ± 4.1	9.2 ± 2.07	39.7 ± 8.6	12.2 ± 1.5	2.42 ± 0.1	475.7 ± 168.8	251.5 ± 22.8	196.3 ± 15.2	555.9 ± 142.7
	Min-max	1.90–5.42	3.54–6.25	71.71–120.9	0.65–2.6	2.58–10.1	6.67–11.55	28.75–47.5	10.5–14.3	2.34–2.55	234.56–610.5	220.88–273.1	176.3–213.3	358.53–698.5
A.c	Mean ± std.	2.93 ± 1.8	4.3 ± 1.05	147.9 ± 25.5	1.46 ± 0.95	10 ± 2.03	25.69 ± 1.8	49.5 ± 5.6	14.5 ± 4.5	4.4 ± 0.5	653.5 ± 119.9	273.8 ± 113.4	171.2 ± 31.3	571.9 ± 68.1
	Min-max	1.35–5.10	3.15–5.3	119.1–175.4	0.78–3.0	7.40–11.9	23.7–27.95	45–57.5	11.2–21	4.10–5.2	564–830.5	117.40–382.2	139.6–214.1	482.46–647.1
S.m	Mean ± std.	2.35 ± 0.8	4.92 ± 1.6	135.7 ± 10.3	1.86 ± 0.7	6.15 ± 2.4	12.6 ± 4.5	47.6 ± 2.4	16.4 ± 2.1	2.01 ± 0.39	593.4 ± 232.2	290.7 ± 27.03	197 ± 15.1	528.5 ± 66.8
	Min-max	1.65–3.3	3.24–6.65	121.2–144.1	1.35–2.9	3.23–8.35	7.33–17.57	45.11–50.5	13.7–18.5	1.56–2.50	256.46–752.5	264.9–319.4	179.8–216.3	472.65–623.4
V.r	Mean ± std.	4.28 ± 1.5	4.47 ± 0.5	340.5 ± 7.8	3.09 ± 0.6	8.95 ± 2.5	10.4 ± 1.18	47.8 ± 1.2	13.8 ± 2	4.18 ± 0.90	443.8 ± 181.4	195.8 ± 66.9	137.4 ± 3.2	209.3 ± 96.9
	Min-max	2.40–5.8	3.93–4.92	330.3–349.1	2.27–3.75	6.34–11.4	9.25–11.47	46.5–49.23	11.8–16.5	3.16–5.32	281.2–618.5	135.24–257.8	134.5–141	123.45–306.1

Abelmoschus esculentum (*A.e*), *Allium cepa* (*A.c*), *Solanum melongena* (*S.m*), *Allium sativum* (*A.s*), *Daucus carota* (*D.c*), *Raphanus sativus* (*R.s*), *Lycopersicum esculentum* (*L.e*), *Vigna radiata* (*V.r*), and *Triticum aestivum* (*T.a*)

of pollution (Nabulo et al., 2012), and agrochemicals (especially metal-based pesticide) is the other contributor of heavy metal pollution in food crops (Lion & Olowoyo, 2013).

12.3.2 Principal Component Analysis for Source Identification

Since the presence of toxic metals in tube well water-irrigated vegetables indicated the possible seepage of element from other contaminated sites, the principal component analysis identified two components accounting for 35.7% and 28% variability of data, respectively. Both components revealed a set of toxic elements from both waste and tube well water sources. In the first component, Pb, Na, Mn, and Mg were dominant. K from wastewater and Na from tube well water had negative association with the component. It seems that major proportion of toxic elements clustered in this component was emerging from industrial and agricultural activities. In second component, the metals detected in wastewater irrigation sources were dominant, including Ni, Cr, Z, Ca, and Fe (−0.69). These elements had strong correlation with this component than those found in tube well water-irrigated vegetable. The clustering of elements into two components showed that the source of toxic elements in the tube well is underground seepage of wastewater. Cr, Co, Ni, and Fe are frequently used in many steel, tannery, and leather industries (Ali et al., 2013). Na and K mostly come from agricultural sites; being parts of fertilizer, they are used in plant supplements (Huang et al., 2007). From such profile, it is more likely that metals were emerging from two distinct sources. The wastewater source of toxic metals may be industrial nonpoint source (Yuan et al., 2013), and the second component shows the agricultural source of heavy metals (Chen et al., 2009). Although both components had complex aggregate of all the elements, either from industrial, agricultural, or residential source, our analysis revealed two distinct conduits were responsible for incorporation of toxic elements into the water bodies (Fig. 12.2). The livestock should not be feed with the fodders that are more metal accumulator. The fodder should be checked and regularly monitored about the contaminant they have and their risk to them.

12.3.3 Human Health and Livestock Risk via Food Crops Consumption

Figure 12.3 summarizes the daily intake of metals for livestock (DIML) in food crops (vegetables and *Triticum aestivum*) grown on wastewater and tube well water. DIM values of buffalo from wastewater and tube well-irrigated food crops followed the trends as Fe>Pb>Mn>Zn>Co>Ni>Cr>Cd>Cu and Fe>Pb>Mn>Zn>Co>Ni>Cu>Cr>Cd, respectively. However, DIM values from

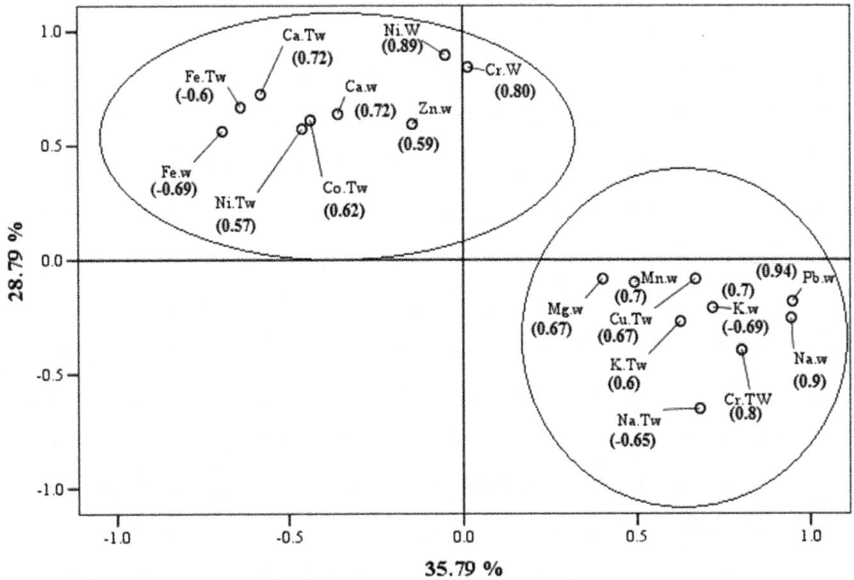

Fig. 12.2 Principal component analysis for source apportionment of heavy metals

wastewater and tube well-irrigated food crops (vegetables and *Triticum aestivum*) in cow followed the order Fe>Pb>Mn>Zn>Co>Ni>Cr>Cd>Cu and Fe>Pb>Mn>Zn>Co>Ni>Cu>Cr>Cd, respectively. DIM values for goats in wastewater-irrigated food crops followed the trends as Fe>Pb>Mn>Zn>Co>Ni>Cr>Cd>Cu, while tube well-irrigated food crops showed the following pattern: Fe>Pb>Mn>Zn>Co>Ni>Cu>Cr>Cd.

The maximum overall daily intake of metals for Fe was recorded in buffalo, cow, and goat through consumption of WWIF and TWIF, while minimum values were observed for Cu in buffalo, cow, and goat due to WWIF and Cd due to TWIF consumption, respectively. Overall DIM in livestock from wastewater-irrigated food crops (WWIF) and tube well water-irrigated food crops (TWIF) consumption were in the order of goat>buffalo>cow. It means that risks were characterized in the same manner to livestock from food crops consumption. Generally, it was observed that highest daily intake of metals for livestock (DIML) was revealed by cereal crop (T.a) as compared to all the vegetables irrigated with wastewater and tube well water.

It was attributed that goat revealed greater DIM values and pose more risks as compared to others because of larger food crops, i.e., vegetables and *Triticum aestivum* consumption. The consumption-to-mass ratio of goat was higher than other livestock members. It also depends on toxicity of metals, e.g., Pb, Cd, Cr, and Ni (nonessential metals) are more toxic than other metals like Fe, Mn, Cu, Zn, and Co (Khan et al., 2013). The DIM in livestock for Fe was higher as compared to other nonessential metals, but the health risk they posed will be less as compared to nonessential metals because they are toxic at very low concentration. This DIM values

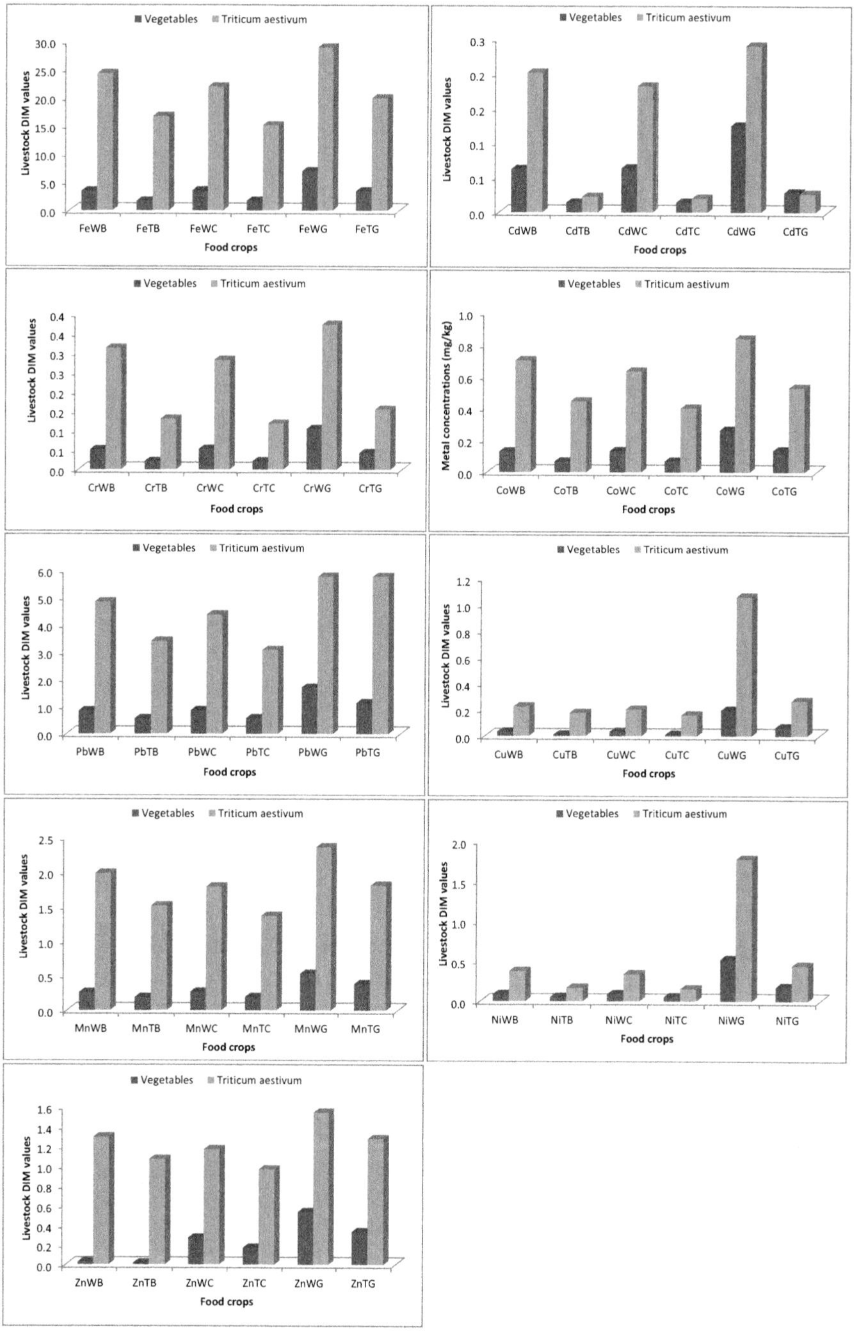

Fig. 12.3 Daily intake of metals (DIM), such as Cu, Zn, Mn, Pb, Fe, Cr, Ni, Cd, and Co, of buffalo (B), cow (C), and goat (G) due to food crops (vegetables and *Triticum aestivum*) consumption irrigated with wastewater (W) and tube well water (T)

for metals in food crops will provide further a baseline for health risk assessment in livestock because RfD values for different metals are present in human and there is an urgent need to provide RfD values for livestock risk assessment.

Wastewater-irrigated food crops showed maximum DIM values as compared to tube well water-irrigated food crops because the wastewater are composite of different metals (Gupta et al., 2012). Livestock consumed the contaminated food crops with heavy metals that become concentrated in the tissue and different organs of livestock which on consumption by human will pose risk to them. These results provide evidence, and good reasons to conclude studying merely edible part of vegetable for heavy metal contents are not enough. This is obvious, that if these vegetables are used in the feed of animals for a longer period of time (which is currently in practice) the animals will no doubt have body burden of these metals which need further investigation on these lines.

12.4 Conclusions

Current study highlighted an aspect which is being neglected, i.e., the risk to livestock and an indirect risk factor for human population via their meat and dairy product consumption. The consumption of Pb-, Cd-, and Cr-contaminated food crops grown on wastewater and tube well water may pose a potential threat to livestock health. High DIM values for livestock were revealed by goat followed by buffalo and cow and posed greater health risk to them due to larger consumption of wheat and vegetables. Higher DIM values for livestock were observed in wheat than vegetables. Results also indicated that heavy metal concentration in edible parts of wastewater-irrigated food crops was much higher than that of tube well water-irrigated food crops. It is suggested that future studies should focus on these aspects and a monitoring of livestock feed reared by human population is also important to estimate the extent of health risk from the consumption of livestock products.

Acknowledgments The authors gratefully acknowledge the contribution of General Lab, the Department of Biochemistry, Dr. Abdul Qadir, and Majid Hussain for their support and guidance during sample collection.

Conflict of Interest The authors declare no conflict of interest.

References

Abbas, S., Sarfraz, M., Mehdi, S., & Hassan, G. (2007). Trace elements accumulation in soil and rice plants irrigated with the contaminated water. *Soil and Tillage Research, 94*, 503–509.

Abbasi, A. M., Iqbal, J., Khan, M. A., & Shah, M. H. (2013). Health risk assessment and multivariate apportionment of trace metals in wild leafy vegetables from Lesser Himalayas, Pakistan. *Ecotoxicology and Environmental Safety*.

Ali, Z., Malik, R. N., & Qadir, A. (2013). Heavy metals distribution and risk assessment in soils affected by tannery effluents. *Chemistry and Ecology*, 1–17.

Al-Othman, Z., & Naushad, M. (2011). Organic–inorganic type composite cation exchanger poly-o-toluidine Zr (IV) tungstate: Preparation, physicochemical characterization and its analytical application in separation of heavy metals. *Chemical Engineering Journal, 172*, 369–375.

Chen, T., Liu, X., Li, X., Zhao, K., Zhang, J., Xu, J., Shi, J., & Dahlgren, R. A. (2009). Heavy metal sources identification and sampling uncertainty analysis in a field-scale vegetable soil of Hangzhou, China. *Environmental Pollution, 157*, 1003–1010.

Gupta, N., Khan, D., & Santra, S. (2012). Heavy metal accumulation in vegetables grown in a long-term wastewater-irrigated agricultural land of tropical India. *Environmental Monitoring and Assessment, 184*, 6673–6682.

Haiyan, W., & Stuanes, A. O. (2003). Heavy metal pollution in air-water-soil-plant system of Zhuzhou City, Hunan Province, China. *Water, Air, and Soil Pollution, 147*, 79–107.

Hashmi, M., Malik, R., & Shahbaz, M. (2013). Heavy metals in eggshells of cattle egret (Bubulcus ibis) and little egret (Egretta garzetta) from the Punjab province, Pakistan. *Ecotoxicology and Environmental Safety, 89*, 158–165.

Hernández-Martínez, R., & Navarro-Blasco, I. (2012). Estimation of dietary intake and content of lead and cadmium in infant cereals marketed in Spain. *Food Control, 26*, 6–14.

Huang, S., Liao, Q., Hua, M., Wu, X., Bi, K., Yan, C., Chen, B., & Zhang, X. (2007). Survey of heavy metal pollution and assessment of agricultural soil in Yangzhong district, Jiangsu Province, China. *Chemosphere, 67*, 2148–2155.

Huang, Z., Pan, X.-D., Wu, P.-G., Han, J.-L., & Chen, Q. (2014). Heavy metals in vegetables and the health risk to population in Zhejiang, China. *Food Control, 36*, 248–252.

Jan, F. A., Ishaq, M., Khan, S., Ihsanullah, I., Ahmad, I., & Shakirullah, M. (2010). A comparative study of human health risks via consumption of food crops grown on wastewater irrigated soil (Peshawar) and relatively clean water irrigated soil (lower Dir). *Journal of Hazardous Materials, 179*, 612–621.

Khan, M. U., Malik, R. N., & Muhammad, S. (2013). Human health risk from Heavy metal via food crops consumption with wastewater irrigation practices in Pakistan. *Chemosphere, 93*, 2230–2238.

Lion, G., & Olowoyo, J. (2013). Population health risk due to dietary intake of toxic heavy metals from Spinacia oleracea harvested from soils collected in and around Tshwane, South Africa. *South African Journal of Botany, 88*, 178–182.

Loutfy, N., Fuerhacker, M., Tundo, P., Raccanelli, S., El Dien, A., & Ahmed, M. T. (2006). Dietary intake of dioxins and dioxin-like PCBs, due to the consumption of dairy products, fish/seafood and meat from Ismailia city, Egypt. *Science of the Total Environment, 370*, 1–8.

Muchuweti, M., Birkett, J., Chinyanga, E., Zvauya, R., Scrimshaw, M. D., & Lester, J. (2006). Heavy metal content of vegetables irrigated with mixtures of wastewater and sewage sludge in Zimbabwe: implications for human health. *Agriculture, Ecosystems & Environment, 112*, 41–48.

Nabulo, G., Black, C., Craigon, J., & Young, S. (2012). Does consumption of leafy vegetables grown in peri-urban agriculture pose a risk to human health? *Environmental Pollution, 162*, 389–398.

Ondo, A. J., Prudent, P., Massiani, C., Biyogo, M. R., Domeizel, M., Rabier, J., & Eba, F. (2013). Impact of urban gardening in equatorial zone on soils and metal transfer to vegetables. *Journal of the Serbian Chemical Society*, 116.

Oves, M., Khan, M. S., Zaidi, A., & Ahmad, E. (2012). Soil contamination, nutritive value, and human health risk assessment of heavy metals: An overview. In *Toxicity of heavy metals to legumes and bioremediation* (pp. 1–27). Springer.

Qadir, A., & Malik, R. N. (2011). Heavy metals in eight edible fish species from two polluted tributaries (Aik and Palkhu) of the River Chenab, Pakistan. *Biological Trace Element Research, 143*, 1524–1540.

Qadir, A., Malik, R. N., & Husain, S. Z. (2008). Spatio-temporal variations in water quality of Nullah Aik-tributary of the river Chenab, Pakistan. *Environmental Monitoring and Assessment, 140*, 43–59.

Rattan, R., Datta, S., Chhonkar, P., Suribabu, K., & Singh, A. (2005). Long-term impact of irrigation with sewage effluents on heavy metal content in soils, crops and groundwater – A case study. *Agriculture, Ecosystems & Environment, 109*, 310–322.

Saha, N., & Zaman, M. (2011). Concentration of selected toxic metals in groundwater and some cereals grown in Shibganj area of Chapai Nawabganj, Rajshahi, Bangladesh. *Current Science (Bangalore), 101*, 427–431.

Singh, A., Sharma, R. K., Agrawal, M., & Marshall, F. M. (2010). Risk assessment of heavy metal toxicity through contaminated vegetables from waste water irrigated area of Varanasi, India. *Tropical Ecology, 51*, 375–387.

Valko, M., Rhodes, C., Moncol, J., Izakovic, M., & Mazur, M. (2006). Free radicals, metals and antioxidants in oxidative stress-induced cancer. *Chemico-Biological Interactions, 160*, 1–40.

Yuan, G.-L., Sun, T.-H., Han, P., & Li, J. (2013). Environmental geochemical mapping and multivariate geostatistical analysis of heavy metals in topsoils of a closed steel smelter: Capital Iron & Steel Factory, Beijing, China. *Journal of Geochemical Exploration*.

Zhuang, P., McBride, M. B., Xia, H., Li, N., & Li, Z. (2009). Health risk from heavy metals via consumption of food crops in the vicinity of Dabaoshan mine, South China. *Science of the Total Environment, 407*, 1551–1561.

Index

H. Alshemmari et al. (eds.), *Contaminated Land and Water*, Emerging Contaminants and Associated Treatment Technologies,
https://doi.org/10.1007/978-3-031-65129-8

P

R

S

T

W

SPRINGER NATURE

GPSR Compliance

The European Union's (EU) General Product Safety Regulation (GPSR) is a set of rules that requires consumer products to be safe and our obligations to ensure this.

If you have any concerns about our products, you can contact us on ProductSafety@springernature.com

In case Publisher is established outside the EU, the EU authorized representative is:

Springer Nature Customer Service Center GmbH
Europaplatz 3
69115 Heidelberg, Germany

The manufacturer's authorised representative in the EU is Springer Nature Customer Service Centre GmbH, Europaplatz 3, 69115 Heidelberg, Germany. If you have any concerns regarding our products, please contact ProductSafety@springernature.com

Printed and bound by CPI Group (UK) Ltd, Croydon, CR0 4YY
30/09/2026
02225604-0016